NIST Technical Note 1726

Development and Use of the NIST Intelligent and Automated Construction Job Site Testbed

Kamel Saidi, Engineering Laboratory

Marek Franaszek, Engineering Laboratory

Jeremy Swerdlow, formerly with Engineering Laboratory

Robert Lipman, Engineering Laboratory

Mani Golparvar-Fard, Virginia Tech

Manu Akula, University of Michigan

Behshad Ghadimi, Virginia Tech

Geraldine Cheok, Engineering Laboratory

Christopher Brown, Engineering Laboratory

Itai Katz, Engineering Laboratory

Paul Goodrum, University of Kentucky

Gabriel Dadi, University of Kentucky

December 2011

U.S. Department of Commerce
John Bryson, Secretary

National Institute of Standards and Technology
Patrick D. Gallagher, Under Secretary for Standards and Technology and Director

DISCLAIMER

Certain commercial entities, equipment, or materials may be identified in this document in order to describe an experimental procedure or concept adequately. Such identification is not intended to imply recommendation or endorsement by the National Institute of Standards and Technology, nor is it intended to imply that the entities, materials, or equipment are necessarily the best available for the purpose.

Abstract

The Intelligent and Automated Construction Job Site (IACJS) Testbed is a large laboratory facility that was established at the National Institute of Standards and Technology (NIST) in 2011. The creation of the testbed was motivated by the need for such testbeds that was identified through several industry initiatives. The capabilities of the testbed include various equipment (such as mobile and cable-driven robots), sensors (such as laser scanners and laser trackers), and computation and visualization resources (such as high-end workstations and high-resolution projectors and cameras).

The application of 3D imaging technologies to six construction industry scenarios, identified in collaboration with industry partners, was selected for potential demonstration in the testbed. The scenario, rebar mapping for cast-in-place concrete railway bridge decks, was selected as the initial construction activity to which 3D imaging would be applied. An industry demonstration was conducted in the testbed on June 30, 2011.

Two 3D imaging technologies, laser scanning and digital images from a camera using custom software to obtain 3D data, were evaluated based on how well they performed and how they affected productivity. The productivity analysis was conducted using Discrete Event Simulation techniques and the ASTM E2204 Standard Guide for Summarizing the Economic Impacts of Building Related Projects (ASTM, 2011).

This report describes the IACJS, the industry demonstration, and results from the experiments conducted during the demonstration.

Contents

List of Tables

List of Figures

1   Introduction

1.1   Motivation and Background

1.1.1   FIATECH Capital Projects Technology Roadmap

Motivation for developing an Automated Construction Testbed (ACT) was based on work conducted in 2003 (Lytle et al., 2003) that was a result of FIATECH's[1] efforts to develop a Capital Projects Technology Roadmap (CPTR[2]), and specifically the Roadmap's Element 4: The Intelligent and Automated Construction Job Site (IACJS). The vision of Element 4 is that the "project site and construction processes of the future will be re-engineered to make use of emerging information and automation technologies to minimize capital facility delivery costs (labor, material and equipment), facility delivery time, and lifecycle costs" (FIATECH, 2003). In order to achieve the above vision the CPTR recommends a strategy that includes seven tasks. Task 6 states (FIATECH, 2003):

> Develop IACJS test bed(s) - This task will focus on the design, development, deployment, and use of an advanced set of test facilities that will provide the first major tests of the IACJS concepts. This test bed will provide realistic environments in a laboratory setting from which proposed and developed IACJS technologies can be tested for potential use on actual construction sites. For example, the RoboCrane was tested extensively within a well-defined and sensed test environment at the National Institute of Standards and Technology (NIST). Those concepts successfully demonstrated through the test bed will then be candidates for deployment on actual construction sites.

The ACT was subsequently renamed the Intelligent and Automated Construction Job Site (IACJS) Testbed in 2006.

1.1.2   NSF-NIST Advanced Building Infrastructure Testbeds (NABIT) Workshop

Impetus to continue work on the IACJS Testbed also came from a 2006 joint National Science Foundation (NSF) and NIST workshop on Advanced Building Infrastructure Testbeds (NABIT), which investigated the need for research validation testbeds (Akin and Garrett, 2006).

---

[1] "An industry consortium that provides global leadership in identifying and accelerating the development, demonstration and deployment of fully integrated and automated technologies to deliver the highest business value throughout the life cycle of all types of capital projects" www.fiatech.org

[2] "A cooperative effort of associations, consortia, government agencies, and industry, working together to accelerate the deployment of emerging and new technologies that will revolutionize the capabilities of the capital projects industry," www.fiatech.org

1

1.1.3 NRC Report Summary and Recommendations

Finally, the National Research Council (NRC) sponsored a workshop held in November 2008 that resulted in a 2009 report which identified five interrelated activities that could lead to breakthrough improvements and advance the efficiency of the capital facility construction industry (National Research Council, 2009). The report represents a consensus view based on discussions with the construction industry of the current challenges to advance the construction industry's productivity and competitiveness. As noted in the report, incorporating these activities into standard practice will require large corporations and government agencies who invest in infrastructure to work together with contractors, architects, and engineers.

The five interrelated activities are:

1. Widespread deployment and use of interoperable technology applications, also called Building Information Modeling (BIM)
2. Improved job-site efficiency through more effective interfacing of people, processes, materials, equipment, and information
3. Greater use of prefabrication, preassembly, modularization, and off-site fabrication (PPMOF) techniques and processes
4. Innovative, widespread use of demonstration installations
5. Effective performance measurement to drive efficiency and support innovation

More specifically, the NRC made three recommendations for NIST to expedite the deployment of the five activities. Recommendation 1 was for NIST and industry stakeholders to develop a strategy to "identify actions needed to fully implement and deploy interoperable technology applications, job-site efficiencies, off-site fabrication processes, demonstration installations, and effective performance measures." As the NRC report suggests, a demonstration installation may take a variety of forms. One of those forms is a scientific laboratory with state-of-the-art equipment and standardized testing and reporting protocols.

Recommendation 2 was that NIST should lead the effort to develop "a technology readiness index to help mitigate the risks of using new technologies, products, and processes by verifying their readiness to be deployed on a widespread basis." Developing and using a readiness index to evaluate technology in the testbed is addressed in a later section on recommendations for future research.

Recommendation 3 was that NIST should "work with the Bureau of Labor Statistics, the U.S. Census Bureau, and construction industry groups to develop effective industry-level measures of tracking the productivity of the construction industry and to enable improved efficiency and competitiveness." Although task-level productivity is discussed in this paper, industry level productivity which is discussed in a NIST special publication (Huang et al., 2009) and white paper (Chapman and Butry, 2008), is not discussed as part of this paper.

1.2 Objectives

The objectives of the IACJS Testbed are to:

1. **Enable innovation**
   The testbed will enable innovation and the development of new technology, processes, and material for the construction industry and academia. To succeed, the testbed must be a collaborative and accessible state-of-the-art environment with easily upgradable systems; be compatible with different computer systems and tools; and allow tool validation, testing, modeling, simulation, and visualization.

2. **Enable evaluation in a controlled testing environment**
   The testbed will enable repeatable testing and evaluation of new construction methods, equipment, and technologies. The testbed environment must be accurately monitored in real time and controlled for construction sensors and processes (real, scaled, and simulated). This environment will assist in the development of best practices and protocols for construction technology and to improve new or existing methods.

3. **Provide an environment for demonstrations**
   The testbed serves as an environment to demonstrate the use of new technologies and processes. These demonstrations will allow for the documentation of the advantages and challenges of technologies and/or processes which will help speed their adoption by industry and drive innovation (by presenting ways to overcome the challenges).

4. **Evaluate technology readiness and mitigate risk**
   The testbed will provide an environment where the readiness of a process, technology, or material can be evaluated to ensure that it is ready to be deployed to the field. Assessing and demonstrating the readiness of a technology to improve a construction process will decrease the risk to companies and provide an understanding of the initial costs versus long term savings.

5. **Enable deployment of technology to the field**
   The testbed will enable the deployment of technology from the research lab to the construction site and ensure that the technology is ready for field use.

1.3 Testbed Requirements

1.3.1 Workshop

In June 2008, NIST held an industry workshop for identifying a set of requirements for the IACJS Testbed. Thirty-five people attended the workshop with 60 % of those from industry and academia and the remaining 40 % from NIST and other research organizations.

The workshop identified four construction activities with the highest priority for investigation in the testbed. These were as follows:

1. Job site management (material management and tracking, safety monitoring, personnel tracking, resource tracking).
2. Capturing existing conditions (capturing time-stamped as-built / existing conditions before, during, and after construction to include temporary and permanent installation).
3. Information availability and accessibility (reliable, repeatable, secure, timely, and efficient transfer of information between producers and consumers).
4. Component assembly and installation (steel, precast, pipe, etc.).

During the 2008 workshop the measurement requirements for some job-site management activities were developed. These requirements are presented in Table 1-1 through Table 1-5. The "Measurement Requirements" column in these tables list the information that must be determined for each of the job site management activities listed in the left-hand side column. For example, for the productivity aspect of the "Personnel Management" activity (Table 1-2), one of the pieces of information needed is the "Idle time" for a crew member. Therefore, the testbed must have the capability to measure worker idle time.

Table 1-1 Material management activity measurement requirements.

Material Management	Measurement Requirements	
Utilization	- Amount received vs. used - Damaged / unusable - Lost / misplaced	- Handling (# of touches) - Re-ordering
Monitoring	- Position in transit - Supply chain status (on-site/off--site/unknown) - Estimated time of arrival (ETA) - Cartesian location	- Cartesian orientation - Grid location - Delivery completeness - Unscheduled deliveries
Identification	- Distinguish/ID material (case pallet, single part) - Material type	- Right material (Bill of material (BOM)/Purchase order (PO)/packing slip/specs) - Quantity

Table 1-2 Personnel management activity measurement requirements.

Personnel Management	Measurement Requirements	
Productivity	- Output - Idle time	- Delays
Monitoring	- On-site / off-site / unknown with time - Zone location - Cartesian position - Body position - Movement	- Current task - Privacy - Crew level - Craft density (# people/area) - Right tools / equipment
Identification	- Craft jurisdiction - Authorization/security - Worker ID - Task assignment	- Certification - Proficiency (capture level of experience (in training, mastered)) - Labor rate - Expected output (RS Means)

Table 1-3 Equipment management activity measurement requirements.

Equipment Management	Measurement Requirements	
Utilization	- Cycle time - Idle time - Delays	- Cost - Breakdown
Monitoring	- On-site/off-site/unknown with time - Zone location - Cartesian position - Cartesian orientation - Configuration - Movement - Current task	- Crew level - Effector / implement location - Emissions - Payload - Health - Energy consumption
Identification	- Technical specifications - Authorization/security - Equipment ID	- Task assignment - Certification for operation

Table 1-4 Site management activity measurement requirements.

Site Management	Measurement Requirements	
Utilities	- Power - Water - Gas	- Sanitary - Communications (internet, cell, global positioning system (GPS), etc.)
Monitoring	- Humidity - Temperature - Pressure	- Lighting - Wind - Precipitation
Identification	- Layout - Map (geographic) - Boundaries - Access (ingress, egress, regress) - Transport	- Location - Accessibility (how hard is it to get to site, adjoining properties) - Security zones - Sub-surface

Table 1-5 Operations management activity measurement requirements.

Operations Management	Measurement Requirements	
Emergency preparedness	- Response time	- Evacuation time
Regulations	- Air quality - Noise - Emissions	- Water quality - Vibration - Injury

2 Description of Testbed Capabilities

2.1 Physical Infrastructure

The IACJS Testbed is located within a laboratory on the NIST campus in Gaithersburg, MD (Figure 2-1).

Figure 2-1 An image of the IACJS Testbed prior to completion.

The laboratory measures nominally 15 m wide, 16 m long, and 10 m high. The floor of the laboratory includes 90 anchorage points each with the capacity of a 900 kN load in the vertical and horizontal directions. The anchorage points are located 1.5 m on center throughout the lab. A roll-up door measuring 5 m wide by 4 m high provides access to the outside. High pressure air and steam as well as vacuum lines are also available in the lab.

2.2 Equipment and Tools

Some of the equipment available in the IACJS Testbed include a six degree-of-freedom (DOF) robotic crane known as RoboCrane™ (Albus et al., 1992) (Figure 2-2); a mobile robot with a 10 kg (22 lb) payload capacity (Figure 2-3); a high-resolution rotation stage with a 454 kg (1000 lb) payload capacity; a pair of high-speed, servo-driven pan-tilt units for positioning sensors; six programmable servo motors and winches; a reconfigurable cable crane apparatus that can be used to test various scaled-down cable-driven robot concepts (Figure 2-4); and a reconfigurable rebar cage (described in more detail in Section 5.3.1).

Figure 2-2 The NIST RoboCrane.

Figure 2-3 Mobile robot.

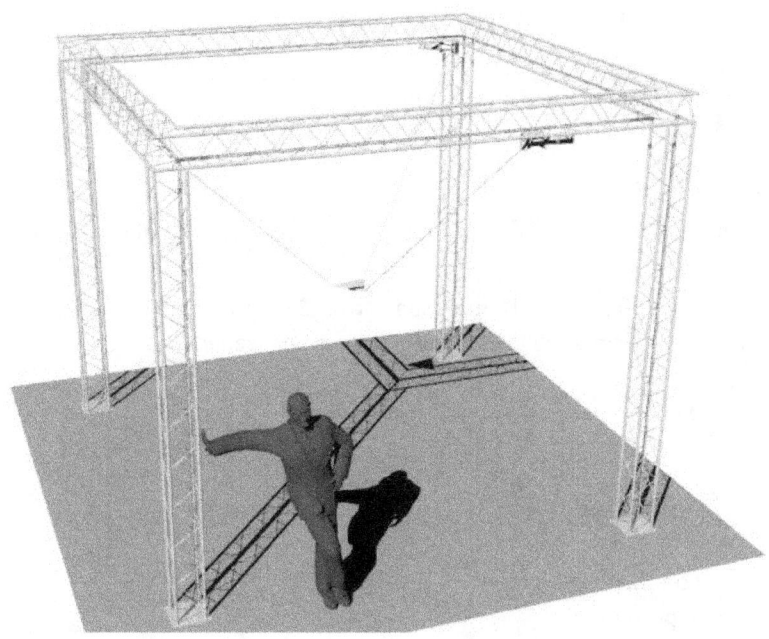

Figure 2-4 3D CAD model of the reconfigurable crane apparatus.

In addition, an overhead gantry crane with a load capacity of 18.14 t (20 ton) spans the entire length of the lab and can lift objects up to a height of approximately 7 m, while a hydraulic pump can supply hydraulic pressure at a rate of 0.0013 m^3/s and a pressure of 34 MPa to any hydraulic equipment.

Finally, the lab is also equipped with a scissor lift with a 9.75 m (32 ft) platform working height and a 318 kg (700 lb) lift capacity.

2.3 Sensors

The IACJS Testbed includes over twelve types of sensors for measuring the 3D shape and static and dynamic locations of objects to within various levels of uncertainty and at different data acquisition rates. The sensors and their key characteristics are presented in Table 2-1. All of the sensors in Table 2-1 are commercial off-the-shelf (COTS) products except for the Calibrated Camera Network (CCN), which uses commercial video cameras and computer hardware along with custom software algorithms. The CCN is described in more detail in Section 2.3.1.

Table 2-1 The IACJS Testbed sensors.

Sensor Name/Type	Key Performance Specifications
Indoor (or Infrared) GPS (iGPS) (Figure 2-5)	± 250 μm error at ≈ 40 Hz update, up to 40 m range
Calibrated camera network (Figure 2-6)	Six pan/tilt/zoom cameras with 1920 x 1080 pixel resolution each
Laser scanners (x5) (Figure 2-7)	From ± 50 μm to ± 10 mm range error, 24 m to 800 m maximum measurement range and up to 900 000 measurements per second
3D range cameras (x7)	
Active RFID (Figure 2-8)	100 m read range, 10 tags, 4 nodes
Total stations (x2) (Figure 2-9)	0.5 arc second angular resolution, ± 0.2 mm range error
GPS (survey grade)	± 0.5 mm position error
UWB ranging radios (x5)	5 W power output
6 DOF laser tracker (Figure 2-10)	± 25 μm 3D position error and ± 3 arc second 3D orientation error, 60 m max range
Dynamic motion measurement camera (Figure 2-11)	± 2 μm resolution at 1000 Hz update, up to 6 m range
Handheld 3D laser scanner (Figure 2-11)	± 60 μm error at >100 Hz update, up to 6 m range
7-axis mobile articulated coordinate measurement machine (CMM) (Figure 2-12)	± 31 μm single point error at up to 2.4 m range
Professional digital camera (x2) and accessories	21 megapixel resolution, 500 mm telephoto lens, 28 mm to 300 mm zoom lens, 45 mm tilt-shift lens, 100 mm macro lens

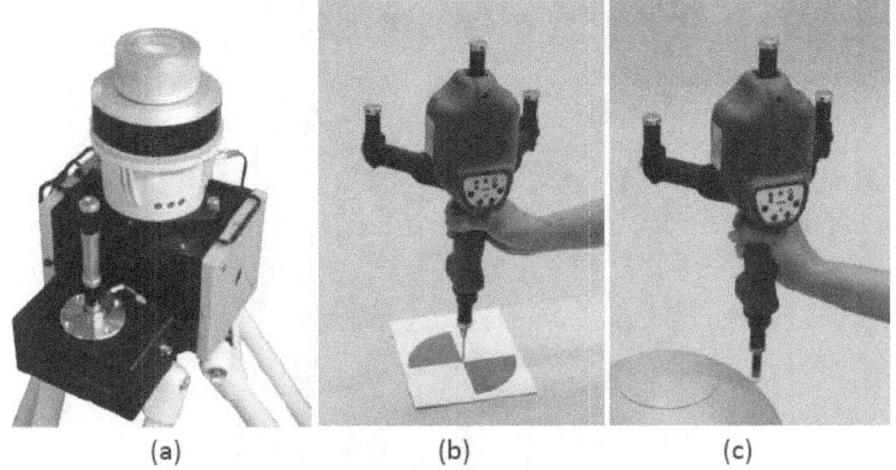

(a)   (b)   (c)

Figure 2-5 Indoor GPS (iGPS) transmitter and mini vector bar (a) and measuring probe with a point tip (b) and a sphere tip (c).

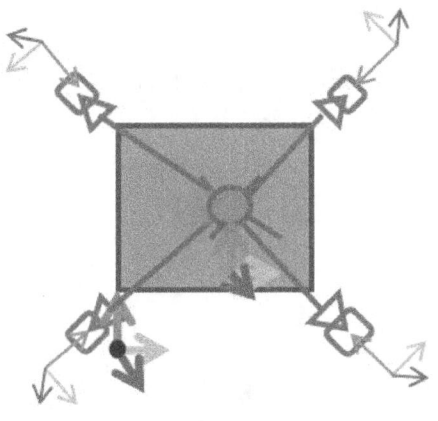

"A calibrated camera network is a collection of three or more cameras, with overlapping views, whose positions and orientations in space are known."

Resolution	1920 x 1080 pixels
FOV (°)	8° (Tele) to 70° (Wide) [f = 3.4 to 33.9 mm]
Pan/Tilt	-100° to +100° (Pan) -25° to +25° (Tilt)

Figure 2-6 The IACJS Testbed calibrated camera network.

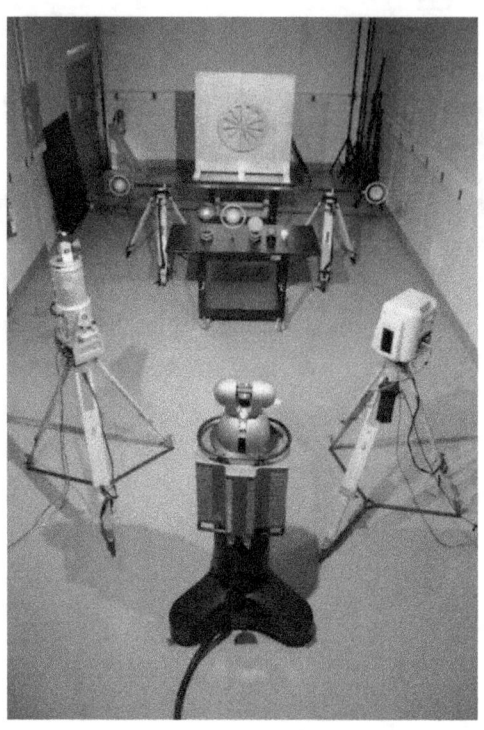

Figure 2-7 A few of the 3D laser scanners available in the testbed.

Proximity
Nodes

Reader
/Writer

Active
Tags

Figure 2-8 The active RFID system.

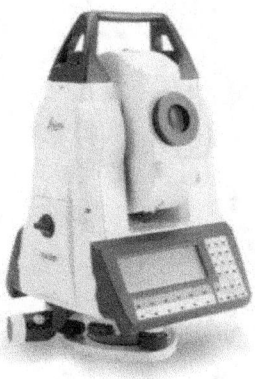

Figure 2-9 An industrial total station available in the testbed.

Figure 2-10 The 6 DOF laser tracker.

Figure 2-11 The dynamic motion measurement camera and handheld 3D laser scanner.

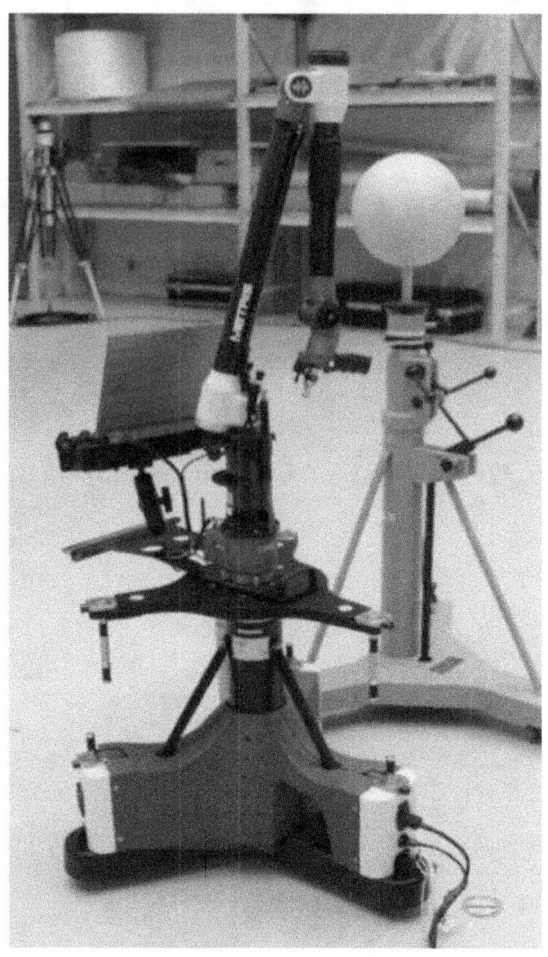

Figure 2-12 The 7-axis mobile articulated coordinate measurement machine (CMM).

2.3.1 Calibrated Camera Network

In addition to high precision laser-based 3D imaging devices, the IACJS Testbed provides a Calibrated Camera Network (CCN) as an alternative 3D sensor. A CCN consists of a collection of cameras whose positions and orientation in space are known. Combining images from these cameras with advanced computer vision algorithms enables applications such as object tracking and 3D scene reconstruction. While CCNs in principle offer similar functionality to laser-based 3D imaging devices, their benefit lies in being highly reconfigurable and at an order of magnitude lower cost. At the present time the technology is still in early developmental stages with few commercial examples. With the rapidly decreasing cost of image sensors, CCNs are becoming an increasingly viable option.

The CCN in the IACJS Testbed consists of a collection of six high resolution HD cameras capable of 1920 x 1080 pixel resolution at 30 frames per second (Figure 2-6). Each camera is connected to a dedicated, network-enabled desktop PC for data collection and distribution. In addition a seventh PC acts as a server to aggregate data and provide a user interface for the system (Figure 2-19). The cameras are rigidly-mounted throughout the lab area to provide maximum overlapping coverage. Each camera is equipped with pan-tilt-zoom (PTZ) capabilities which allows the work volume to be modified as per experimental requirements.

To operate the CCN as a sensor for object tracking, several practical details must be considered; these are described in the following three sections.

2.3.1.1 Calibration

Calibration in this context refers to the estimation of camera parameters needed to map a point in the world reference frame (WRF) to a pixel on the image sensor. As per the perspective projection model, the calibration is defined as the product of two matrices (see Figure 2-13). The first matrix, containing the *extrinsic parameters*, describes the mapping of a 3D coordinate in WRF to a 3D coordinate in the local camera reference frame (CRF). This extrinsic matrix includes the rotation and translation (i.e., position and orientation) of the camera in space. The second matrix, containing the *intrinsic parameters*, maps the 3D point in CRF to the discretized pixel grid. These parameters primarily consist of the focal length and image center of the lens/CCD stack.

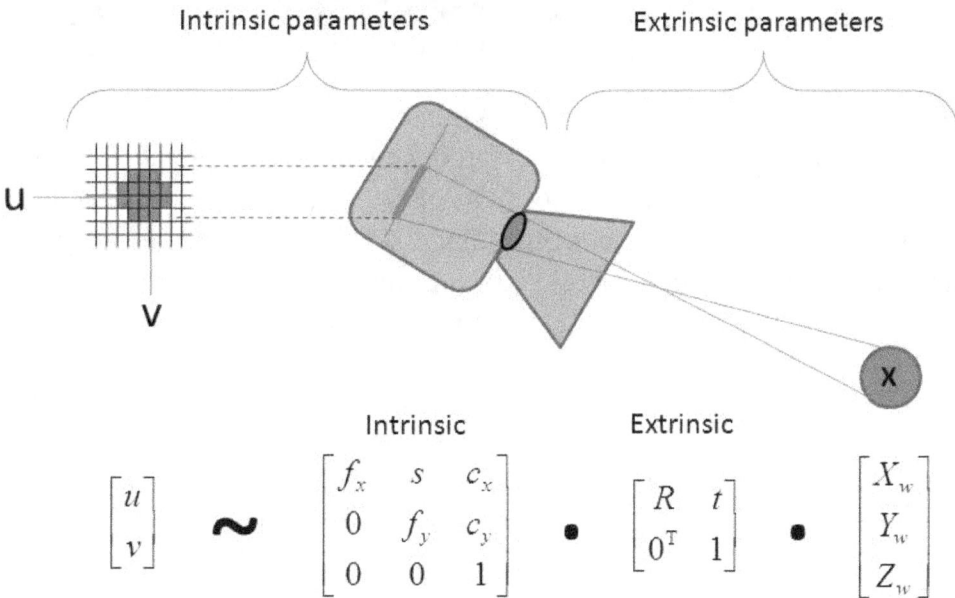

$$
\begin{bmatrix} u \\ v \end{bmatrix} \sim \begin{bmatrix} f_x & s & c_x \\ 0 & f_y & c_y \\ 0 & 0 & 1 \end{bmatrix} \bullet \begin{bmatrix} R & t \\ 0^T & 1 \end{bmatrix} \bullet \begin{bmatrix} X_w \\ Y_w \\ Z_w \end{bmatrix}
$$

Figure 2-13 The CCN perspective projection model.

In the demonstration, intrinsic parameters were estimated using the Camera Calibration Toolbox (Bouguet, 2010) in MATLAB. The toolbox is based on the algorithm described in (Zhang, 1998), which estimates parameters from a sequence of images of a checkerboard pattern. This package also provides distortion parameters to compensate for barrel distortion induced by the lens (see Figure 2-14).

Figure 2-14 Radial lens distortion before and after correction with intrinsic parameters.

For the demonstration, ten test spheres distributed throughout the work volume were used as calibration targets (see Figure 2-15). The 3D positions of the spheres were measured using the iGPS system (see Figure 2-5c), while the corresponding 2D pixel positions were identified manually in each image from the four cameras that were calibrated.

Figure 2-15 The view (from one of the CCN cameras) of the ten CCN calibration spheres in the IACJS Testbed.

2.3.1.2 *3D Point Reconstruction*

While the calibration parameters map 3D world coordinates to 2D pixel coordinates, the primary application of the CCN is the inverse problem: *Given a series of images containing an object of interest, what is the object's position (and possibly orientation)?* The general approach involves solving an optimization problem as depicted in Figure 2-16. While the mathematical details are beyond the scope of this report, reconstruction is covered extensively in Hartley and Zisserman, 2000.

Projection matrices P_i contain the calibration parameters as described above. The camera pixel coordinates (u_i, v_i) containing the object of interest are generally obtained through standard object recognition algorithms. In the demonstration, a large, static test sphere (see Figure 2-17) was manually identified in each image.

17

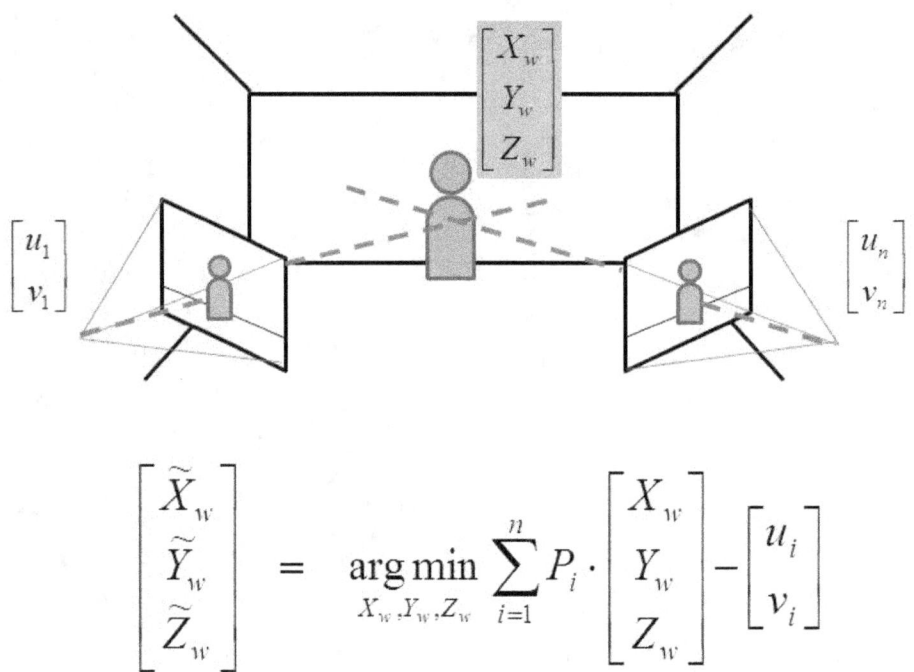

$$\begin{bmatrix} \tilde{X}_w \\ \tilde{Y}_w \\ \tilde{Z}_w \end{bmatrix} = \mathop{\arg\min}_{X_w, Y_w, Z_w} \sum_{i=1}^{n} P_i \cdot \begin{bmatrix} X_w \\ Y_w \\ Z_w \end{bmatrix} - \begin{bmatrix} u_i \\ v_i \end{bmatrix}$$

Figure 2-16 Solving the 3D point reconstruction problem, given pixel coordinates (u,v) and camera matrices P.

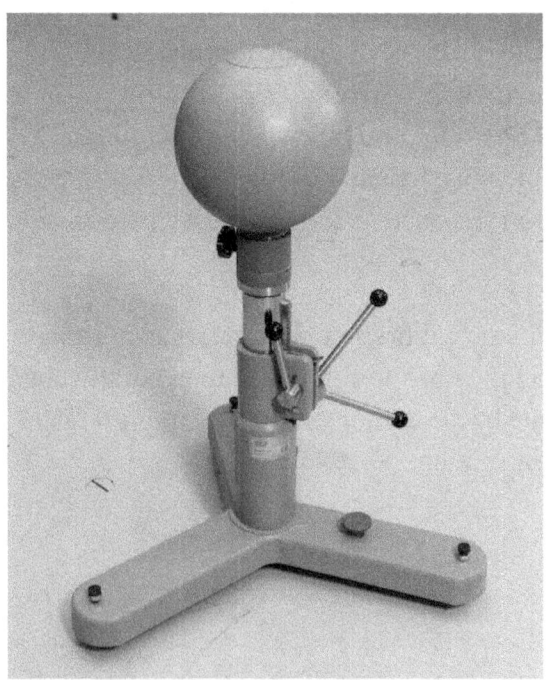

Figure 2-17 The large test sphere used to evaluate the static performance of the CCN.

2.3.1.3 Triggering and Synchronization

Triggering refers to the ability for multiple cameras to make simultaneous observations by means of a common signal. When cameras are triggered, they acquire images at nearly the same instant, usually within a few milliseconds of each other. This is a critical consideration for tracking dynamic targets using a CCN, as a discrepancy of only a few tens of milliseconds can result in a localization error on the order of centimeters for objects moving at pedestrian speeds. High-precision triggering is accomplished by means of a signal generator and a dedicated hardware port on the camera. This is typically only found on higher-end camera units. For low-cost devices such as webcams, this process must be accomplished through software and is dependent on a host computer being specially provisioned to meet real-time constraints. Triggering is obviously a very desirable property but is surprisingly difficult to achieve in practice. How the bus is polled, how the CPU schedules threads, and clock skew resulting from long cable runs to the cameras all affect triggering in ways that are hard to predict. This can be mitigated with the use of one or more signal generators that provide a clock signal to each camera (precluding the possibility of an entirely wireless system and potentially increasing system cost significantly).

Significantly easier to achieve is synchronization, where all elements maintain the same clock (through network time protocol, for instance), but are free to take observations independently. With an accurate time stamp, latency can be taken into account when designing algorithms specific to camera networks, reducing the requirement for adequate triggering.

While triggering and synchronizations are expected to be incorporated into the IACJS Testbed in the near future, no such provision was necessary given the static nature of the demonstration.

2.4 Computing and Visualization

2.4.1   Computation and Hardware Resources

The testbed includes two desktop workstations (Figure 2-18) with dual 8-core processors, 48 GB of RAM, 5 GB of video RAM with 240 compute unified device architecture (CUDA) parallel processing cores and 3 TB of hard disk storage; as well as a mobile workstation with a quad-core processor, 16 GB of RAM, 1 GB of video RAM, 1 TB of hard disk storage and a 432 mm (17 in) red green blue (RGB) LED display.

19

Figure 2-18 Desktop (left) and mobile (right) workstations.

In addition to the above computing capabilities a ruggedized notebook computer as well as two tablet computers are also available in the testbed.

2.4.2 Software Resources

Over a dozen Computer Aided Design / Computer Aided Manufacturing (CAD/CAM) and point cloud processing software applications are available for use in the IACJS Testbed. In addition, access to various numerical/statistical analysis, image processing, and finite element analysis software is also available.

2.4.3 Visualization and Video Recording Capabilities

Installed within the testbed are a large screen display (52 in LCD HDTV) as well as a 3 LCD projection system (4000 ANSI Lumens, 1400 x 1050 pixel resolution). The HDTV is mounted on top of a rolling computer desk (see Figure 2-18) while the projector is mounted in the laboratory's ceiling in order to project images onto the floor.

In addition, HD movie capture is possible through the six HD pan/tilt/zoom cameras - four mounted in each corner of the lab at a height of 7 m and two mounted in the ceiling – that make

20

up the calibrated camera network (CCN) described in Section 2.3.1. Each camera is connected to a dedicated computer located on a rolling desk with a bank of monitors (Figure 2-19). A seventh computer (and monitor) is used as the control center for the CCN. The cameras can all be controlled through a remote control unit located on the same desk.

Figure 2-19 The HD camera computers, monitors, and remote control unit on a rolling desk.

3 Identification of Industry Scenarios

Upon completion of the construction and instrumentation of the IACJS Testbed in April 2011, the capabilities of the testbed were demonstrated through characterizing the performance of technologies that aim to solve an industry problem for a particular infrastructure construction scenario.

3.1 Scenario Selection Method

Potential infrastructure construction scenarios were identified through examining infrastructure projects with work processes that could benefit from the use of advanced technology, advanced project controls, or building information modeling.

Based on a review and discussion of current construction projects and construction tasks as identified by the authors and industry professionals, a group of six example construction scenarios was identified. Specific tasks were identified from these scenarios that would require the use of the testbed. Tasks that were crosscutting, i.e., spanning multiple industry sectors, were preferred so that process improvements might have the widest impact.

3.2 Potential Scenarios

Six infrastructure construction scenarios and crosscutting challenges were identified and are listed in Table 3-1. The right-most column in Table 3-1 lists potential construction sectors to which the challenges are applicable. Each scenario in Table 3-1 is explained in more detail in the subsequent six sections.

Table 3-1 Cross-cutting themes for six testbed scenarios.

Section	Scenario	Cross-Cutting Challenge	Construction Sectors
3.2.1	Placement of Embedments for Railway Pavement	Recording rebar location before pouring concrete	Bridge and rail decks, nuclear plants
3.2.2	Placement of Bridge Structural Components	Placement of prefabricated structural components	Bridge deck and piers, commercial and industrial building components
3.2.3	Nuclear Power Plant Containment Vessel Ring/Dome Placement and Fitting	Modular fit with mating structures	Components for nuclear plants, bridges, and commercial and industrial buildings
3.2.4	Inspection of Nuclear Power Plant Pressure Vessel for Construction	Verify shape and dimensions of critical structural parts	Components for nuclear plants, bridges, and commercial and industrial buildings
3.2.5	Building Structural Steel Erection	Measuring plumbness of structural parts	Components for nuclear plants, bridges, and commercial and industrial buildings
3.2.6	Critical Crane Lifts Where the Crane Operator's View of the Load is Limited	Critical crane lifts with limited visibility	Nuclear plants, buildings, bridges

3.2.1 Placement of Embedments for Railway Pavement

The construction of a railway line requires embedments in the steel reinforced concrete pavement underneath the length of railway tracks. Not knowing where the rebar cages are placed within the concrete makes drilling into the concrete risky in terms of worker safety and structural integrity of the pavement. Therefore, the drill bit must avoid contacting the rebar. Although negatives (such as wooden dowels) may be placed in the rebar cage prior to pouring the concrete to create the voids for the embedments, this practice is labor intensive and fraught with problems. A better method would be to map the locations of the rebar prior to concrete pouring.

Measuring the rebar locations quickly and accurately prior to pouring concrete for railway bridge concrete decks could increase the productivity of the rebar placement, concrete pouring, and finishing activities. It would eliminate the steps required to place wooden dowels (or other negative shapes) within the rebar, improve access to the rebar during concrete pouring, and may allow the use of concrete finishing machines. In addition, maps of the rebar locations would be a valuable tool for future retrofit or rehabilitation of the railway decks.

3.2.2 Placement of Bridge Structural Components

This scenario is based on the replacement of an existing bridge that spans a primary highway. Pre-fabricated bridge piers were chosen to reduce the construction schedule and reduce the impact to the traveling public. The alignment and placement of pre-fabricated bridge piers into the supporting barrier wall and ensuring that there are sufficient clearances between the rebar extending from the bottom of the pier, the rebar of the barrier wall, and the footing rebar is a critical process. The original design for the piers used J-bars (rebar shaped like the letter "J") that extended 1.8 m (6 ft) below the pier which were intended to fit between the supporting barrier wall rebar. Due to the rebar complexity, narrow clearances, and the possibility of rebar misalignment and entanglement, the construction contractor decided to mitigate these risks by changing the rebar configuration to 0.3 m (1 ft) straight rebar with the remaining J-bar length attached with couplers on-site after the piers were set. Given a choice of technological tools, the alignment of the rebar may have been confirmed such that the J-bars could have been cast into the pre-fabricated piers saving time and reducing cost.

This process is a crosscutting challenge for any construction field where rapid, inexpensive capture of 3D spatial data for confirming exact rebar location, clearance, and tolerance compliance is important. The application of advanced measurement technologies could reduce project duration and labor intensive tasks, and advance the adoption of prefabrication and accelerated bridge construction.

3.2.3 Nuclear Power Plant Containment Vessel Ring/Dome Placement and Fitting

New nuclear power plants (NPP) use modular designs in order to improve quality, efficiency, and cost. These designs require that the reactor containment vessel be manufactured in sections that are shipped to the site for assembly. Each assembled containment vessel ring can be over 40 m in diameter and must fit on top of rings already installed and accommodate subsequent rings. Placement and fitting of the containment vessel rings and domes into their final assembled locations is a time consuming and costly task since it requires that the crane hold the ring in place while the alignment is manually verified and any needed adjustments are performed. Checking the alignment and making the adjustments requires that workers and any needed equipment be placed on elevated platforms or scaffolding at the required ring height. If the dimensions of the assembled rings and domes could be verified and adjusted on the ground before they are lifted into place, the potential time cost savings could be significant considering that each containment vessel may be made of eight to nine rings.

There are several potential benefits to verifying how containment vessel modules will fit-up with mating structures before they are lifted into place such as early detection of problems which will allow for remedial actions to be taken and re-scheduling of personnel and equipment as needed, reducing fit-up time, crane usage, and risk.

25

3.2.4 Inspection of Nuclear Power Plant Pressure Vessel for Construction

The manufacture of the liner for the containment vessel for NPPs must meet strict inspection standards as specified in ASME (American Society of Mechanical Engineers), Section III, Division 2, Boiler and Pressure Vessel Code. The liners are usually manufactured from several pieces of steel plate rings or dome sections that are welded together into the final vessel either at the manufacturing facility or on-site. In order to meet the dimensional tolerances for the overall vessel, the individual parts often have to be manufactured to higher tolerances. Verifying the dimensions of the individual parts and of the assembled vessels is a labor intensive and time consuming process. The current methods involve constructing templates that are placed against the surface of the vessels and then manually measuring the gaps between the templates and the walls of the vessel.

Potential benefits for using 3D imaging to verify the shape and dimensions of the individual parts and of the assembled pressure vessel modules include significantly reducing inspection time, increasing inspection data quality, and reducing the time and cost for creating final as-built models.

3.2.5 Building Structural Steel Erection

During the steel erection process for building construction, the steel columns must be checked for plumbness. One practice uses a total station and a hand held retro-reflector target that must be placed on the steel surface. Another uses a reflectorless total station to take measurements along the surface of a column. Plumbness may be more efficiently and accurately determined using different sensing technology. This process is a cross-cutting challenge for any construction field where rapid, inexpensive capture of 3D spatial data for confirming steel plumbness, clearance, and tolerance compliance is important. The benefits of using 3D imaging to measure plumbness include reducing the survey crew's field time, allowing multiple columns to be measured with one scan, and providing data for as-built documentation.

3.2.6 Critical Crane Lifts Where the Crane Operator's View of the Load is Limited

During many critical lifts (such as when a reactor or other structure is being lifted into its final position inside the containment vessel) crane operators have limited or no visibility of loads during the crucial parts of the lift. The visibility is often impeded by obstacles or by the structure into which the load is being lowered. In addition, the crane operator must constantly be aware of the position and orientation of the crane in relation to hazards within the crane's environment (such as overhead utility lines). Although critical lifts are planned in advance and all hazards and obstacles are identified, the crane operator often relies completely on radio commands from a spotter. This slows down the process since the operator is essentially "flying blind" and has to rely solely on the spotter to be alert for any unforeseen obstacles or hazards that may occur at the time of the lift.

Providing situational awareness to a crane operator in the form of real-time 3D information about the position of the load relative to structures, obstacles, and hazards would be beneficial. Enhancing the crane operator's situational awareness can potentially improve both critical lift productivity and safety. A secondary benefit would be a capability to log crane and load motions in 3D space relative to their surroundings for productivity analysis, training, and for optimizing future lifts (Schoonmaker, 2010).

3.3   Selected Scenario for Demonstration

3.3.1   Selection Process

Once the six scenarios were identified, a meeting was held on November 29, 2010 with industry and academic professionals to review the six scenarios and provide feedback as to the best scenario to choose as the pilot study in the testbed. The scenario, Placement of Embedments for Railway Pavement, was chosen as the pilot scenario for its relevance to a current construction project located in close proximity to NIST, which allowed on-site visits and experiments. This scenario was also chosen because the process of mapping rebar could be an important task in several industry sectors including nuclear plant construction.

3.3.2   Description of the Rebar Mapping Scenario

3.3.2.1   Project Background

As mentioned previously, the construction of a railway line at times requires embedments in the reinforced concrete pavement underneath the length of the railway tracks. For the project with this specific scenario, the embedments are located where the plinth stirrups will be anchored into the concrete for attachment to the railway tracks. When the holes are drilled in the concrete pavement, contact with the rebar must be avoided. Drilling into the steel reinforcement is a safety hazard for the operator, compromises the structural integrity of the pavement, introduces an initiation point for reinforcement corrosion, and damages the drill. Although custom made dowels (or blockouts) may be placed into the rebar cage prior to pouring the concrete, the practice of removing dowels after concrete placement is labor intensive. An alternative method is to digitally map the locations of the rebar voids for embedment locations prior to concrete placement.

3.3.2.2   The Original Method

The existing design specified the use of 3.8 cm (1.5 in) diameter wooden dowels as blockouts to mark the embedment locations in the rebar cage every 0.38 m (15 in) (Figure 3-1). The blockouts were screwed into the bottom of wooden planks to hold them in place. The wooden planks were tied to chairs attached to the top layer of rebar (Figure 3-2). The wooden planks were also necessary to create a recess along the length of the pavement. The top surface of the wooden planks matched the final grade of the concrete. After the concrete was placed and had set for a few days, the wooden planks and dowels were removed. The plinth stirrups were anchored into the voids left by the dowels with epoxy.

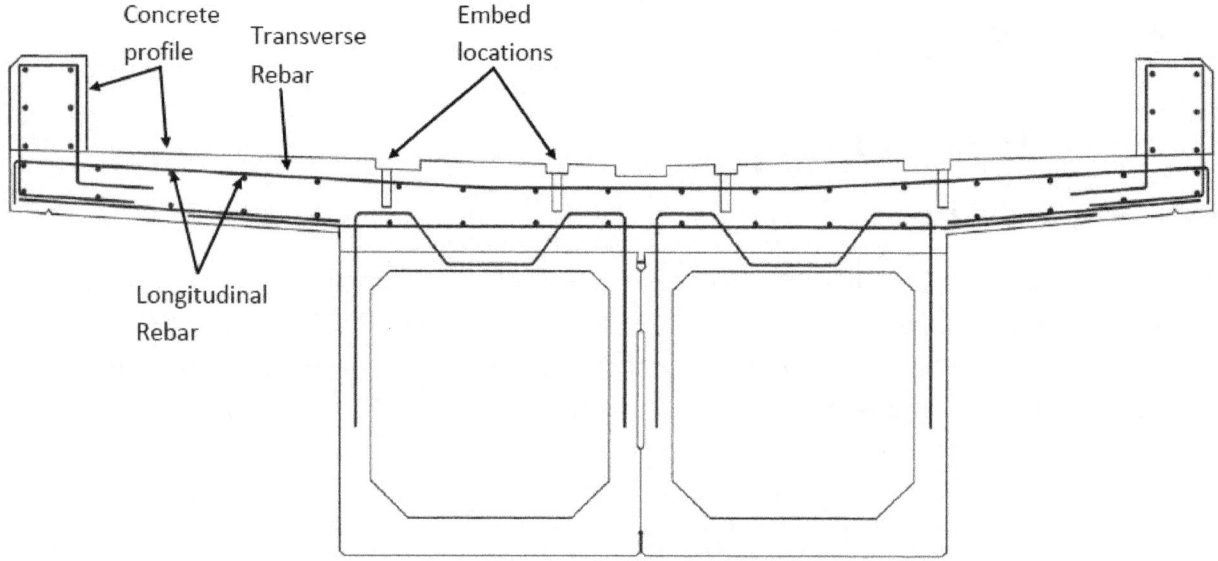

Figure 3-1 A cross-section of the railway bridge deck.

Figure 3-2 A picture of the railway bridge deck with the dowels and wooden planks installed.

The addition of the wooden dowels and planks increased the congestion in the rebar mats and could adversely affect the quality of the concrete by creating honeycombs and voids. Additionally, it was more labor intensive as it restricted access and movement of the workers placing, vibrating, and finishing the concrete. The removal of the dowels involved locating the wood planks as they were often covered with concrete. Once located, removing the dowels required two drill bits, one wood bit to remove most of the dowel, and a second masonry bit with a slightly larger diameter to remove the remaining dowel. Each hole was then covered with a foam plug to prevent debris from entering and to protect it from any damage (Figure 3-3).

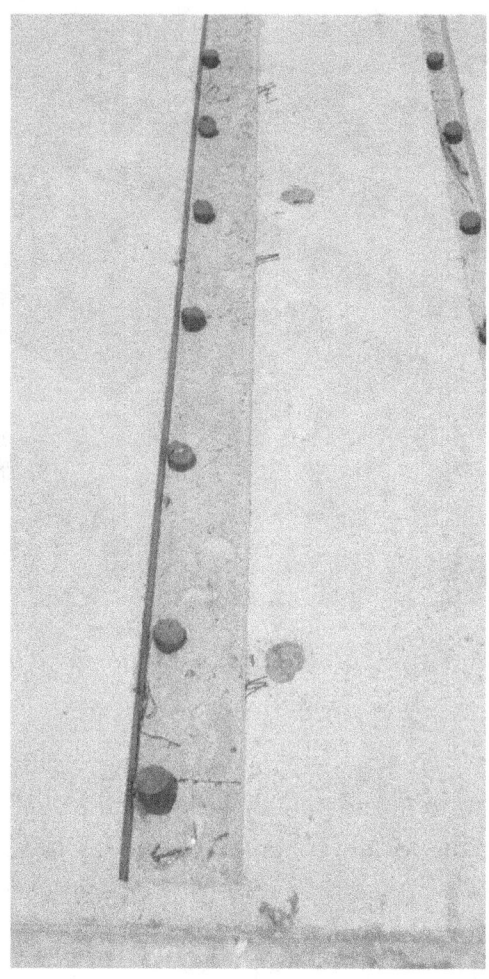

Figure 3-3 Embedment holes with blockouts removed and foam plugs inserted.

3.3.2.3 An Alternative Method

The alternative method, to digitally map the rebar layers including the rebar-free spaces, would provide the ability to measure and mark the locations for the embedments after the concrete is placed. This method eliminates the time and cost required to place the wooden dowels within the steel reinforcement mats and remove them from the cured concrete. It would improve access during concrete placement by eliminating obstacles to the flow of concrete. In addition, digital maps of the reinforcement locations would be a valuable tool for future retrofit or repair of the railway decks. This method also eliminates manual activities that are associated with wood blockouts in the original method. The added steps involved are generally described as defining a site coordinate system (SCS), measuring targets in the SCS, setting up the equipment, measuring registration targets, acquiring data, moving to the next location and repeating the previous three steps, processing the data, and finding and drilling out the locations for the embedments.

The alternative method could significantly improve production for the concrete deck placement operation, avoiding the time and cost to place and remove dowels, and shorten the project duration.

4 Evaluation Metrics

In order to characterize the measurement performance of the different technologies, performance metrics were developed. In addition to performance metrics, a technology may also be characterized based on how it affects productivity – productivity metrics. Generally, there is an existing/current method to perform a certain task. The difference in the productivity between the status quo and the alternative method (e.g., using a new technology) will allow for comparisons between the two methods.

This chapter describes the performance and productivity metrics used for the selected scenario.

4.1 Performance Metrics

The technologies evaluated for the selected rebar scenario should provide the location and depth for safe drilling. This requirement implies that the technology has the capability to acquire 3D measurements with accuracy appropriate for the particular application and to determine the status (safe or unsafe) of a potential drilling location. Thus, the performance metric should report on two aspects of the technology being evaluated:

1. spatial deviation of the measured points from ground truth, and

2. accuracy of determining where it is safe or unsafe to drill in a location – based on the 3D data obtained and the classifier.

The performance metrics were developed based on the following assumptions:

1. Drilling occurs on a flat planar surface. This may be either a vertical wall or horizontal/sloped floor slab. The data processing method would have to be modified for curved or arched surfaces.

2. The drilling plane is parallel to the base plane which supports the rebar cage (e.g., bottom surface of the slab) or to the opposite surface of a wall for a vertical wall, see Figure 4-1.

3. Drilling is done in a direction perpendicular to the drilling plane.

4. The rebar mat closest to the drilling plane is used to form a grid and safe drilling locations are inside quadrilateral cells with vertices defined by the corresponding rebar intersections

5. Each cell is either safe or unsafe for drilling. Although the drilling depth is calculated for each cell, "safe" status is assigned only when there is no obstacle between the drilling

plane and the base plane anywhere within the entire cell. In all other cases, the cell status is classified as unsafe.

Based on the above assumptions, the measurands are the coordinates of the set of points, rebar intersections, projected onto the base plane. The ground truth coordinates were based on the rebar intersections as obtained in Section 5.3.2.1.

The ground truth for the cell status (i.e., safe or unsafe) was determined visually. Based on these definitions of ground truth, the following metrics were used to gauge the measurement performance of the technologies being evaluated:

1) the mean deviation of the projected rebar intersections from ground truth, and

2) the number of false positives of cells classified as safe and the number of false negatives of cells classified as unsafe.

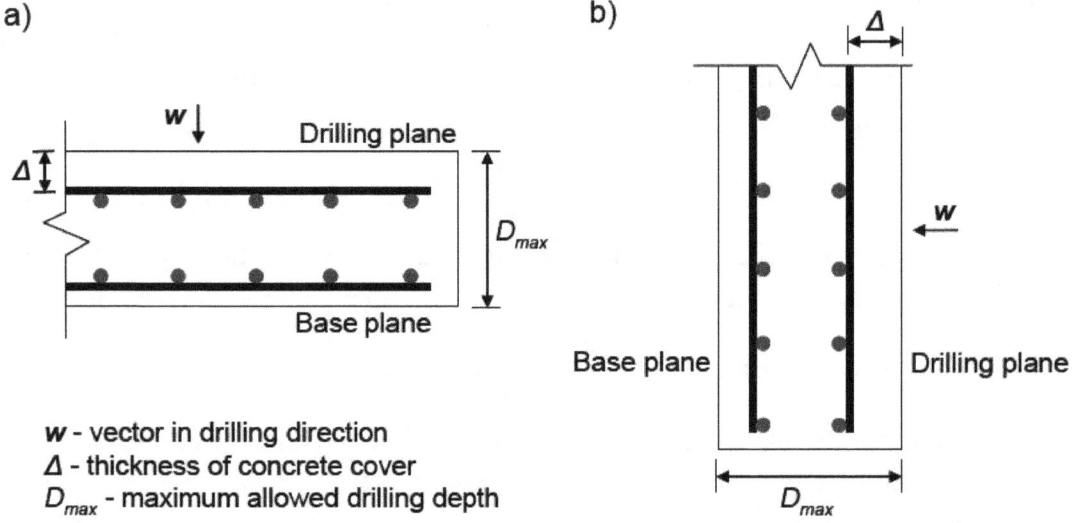

Figure 4-1 Schematic showing the drilling plane and base plane for reinforced concrete slab (a) and vertical reinforced wall (b).

4.2 Productivity Metrics Using ASTM E 2204 and Simulation

The ASTM E2204 Standard Guide for Summarizing the Economic Impacts of Building-Related Projects (Standard Guide for Summarizing the Economic Impacts of Building-Related Projects) (ASTM, 2011) can be used to compare multiple construction methods. The summary format consists of three sections: a description of the significance of the project, the analysis strategy,

34

and the calculation of benefits, costs, and other key measures. The guide is applicable to building and non-building related projects and research related to those projects.

Specifically for the application in the testbed, this guide can be applied to summarize the economic performance of different construction methods using different technology. It can serve as an initial evaluation tool to predict cost and time and assist in determining which alternative methods should be considered in the testbed.

Once a construction task has been identified for study in the testbed, then the available technology choices that can be considered to apply to the task are defined. The alternative methods using different tools are determined (including the assumptions made). The E2204 standard can then be applied to predict the benefits, costs, and additional measures for the existing and alternative methods. Based on the economic performance, the best method (or methods) is then chosen for further study in the testbed (Figure 4-2).

Economic Performance: ASTM Standard E2204

E2204 is an ASTM Standard designed for summarizing the economic performance
of alternative construction-related procedures and technologies

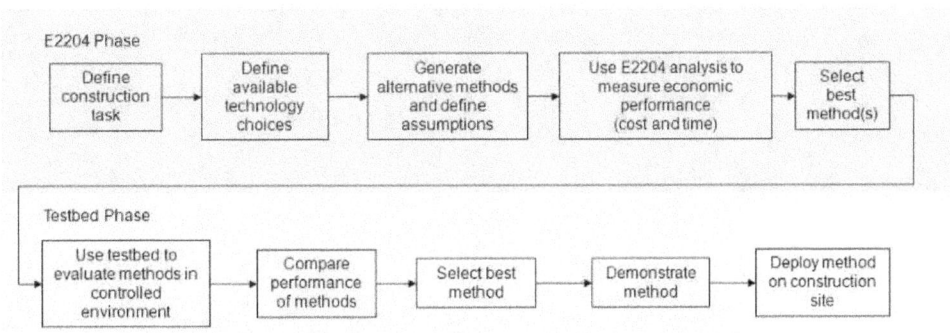

Follow up to compare actual use on site to theoretical use
(E2204 estimates and testbed measurements)

Figure 4-2 Using the ASTM E2204 standard to identify alternative methods for use in the testbed.

5 Demonstration of the Selected Scenario in the IACJS Testbed

This section describes how the IACJS Testbed was used to characterize and to compare various technologies for mapping rebar locations within a railway bridge deck before concrete is poured. This capability was demonstrated in the testbed to an audience from industry and academia on June 30, 2011.

5.1 Demonstration Architecture

The basic architecture for the demonstration involved sensing the state of the rebar with different types of sensors and then processing the data using commercial and custom software algorithms, which result in a characterization and/or comparison of the various sensors used (Figure 5-1).

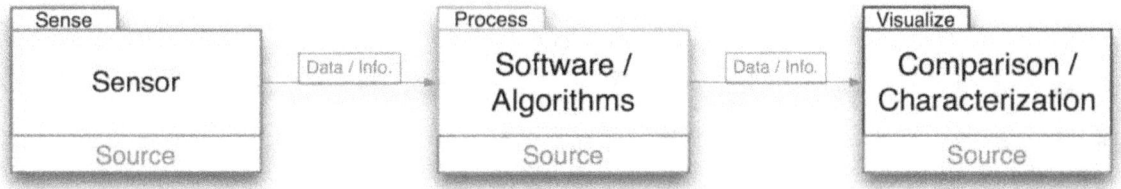

Figure 5-1 The basic testbed architecture used for the June 30, 2011 demonstration.

Figure 5-2 shows a more detailed architecture diagram of the June 30, 2011 demonstration. As seen in this diagram, two sensors, the infrared (or indoor) GPS (iGPS) and the Coherent Laser Radar (CLR) (hereafter considered the "ground truth" 3D imaging instrument), provided information about where the rebar and the sensors were (i.e., the "ground truth" data). The iGPS system was also used to register the CLR, CCN, laser scanner, and photogrammetry data to a common coordinate system.

Three dimensional (3D) point clouds of the rebar were obtained using[3] 1) a digital camera and photogrammetry software algorithms and 2) a 3D imaging system (a laser scanner). In addition, a ground truth point cloud of the rebar was also obtained using the CLR (with a manufacturer specified uncertainty of ± 100 μm). All three point clouds were processed through the same

---

[3] Note that the CCN was not used to map the rebar due to lack of time.

custom-written algorithm (see Section 6.3) in order to detect the spaces in between the rebar (the rebar-free spaces) in which it is safe to drill after the concrete is poured. The processing of the data is explained in further detail in Chapter 6.

Finally, the processed data were compared to the ground truth data and visualized using three separate techniques: 1) building information modeling (BIM); 2) a laser projector; and 3) a drilling simulation. These techniques are explained in sections 5.3.8.1, 5.3.8.2, and 5.3.8.3, respectively.

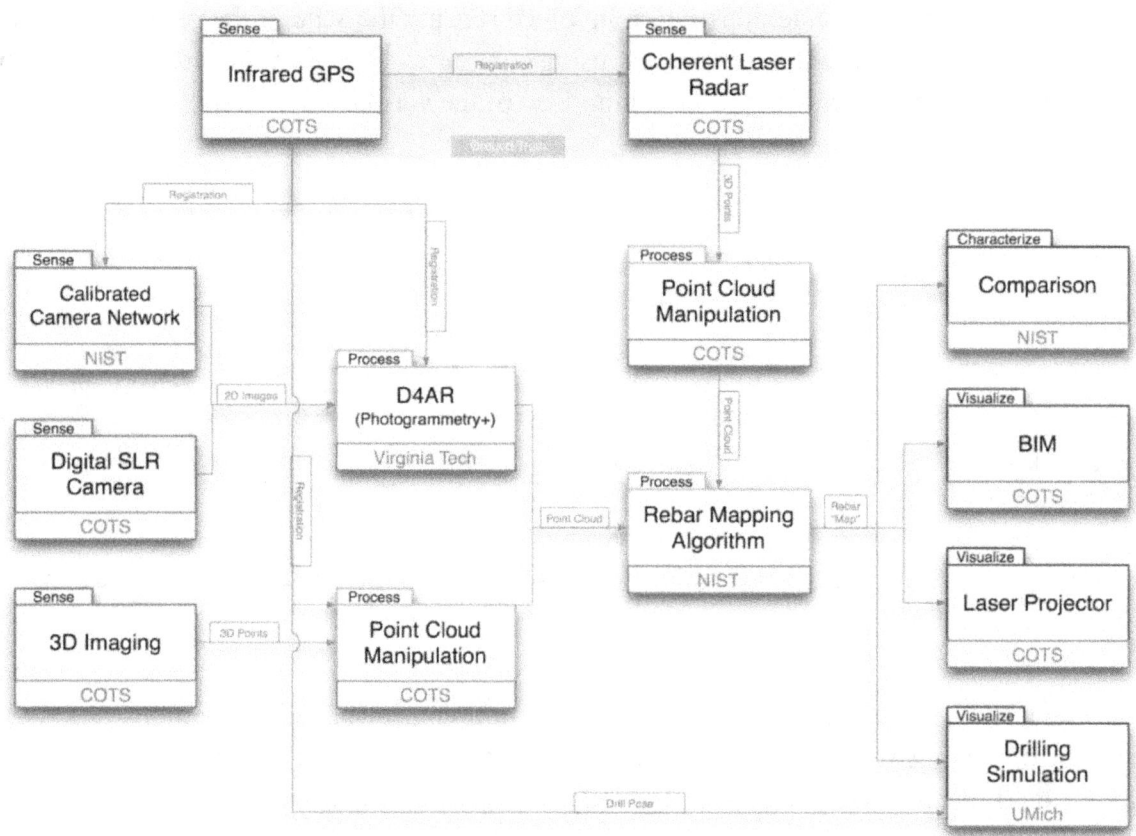

Figure 5-2 The detailed architecture diagram for the June 30, 2011 testbed demonstration.

5.2   Selected Technologies

Methods to locate rebar in concrete fall into two general categories that are differentiated from each other by when they are applied. One method may be used to map the rebar before the concrete is poured (the pre-concrete method) while the other method may be used after the concrete is poured and has cured (the post-concrete method). The choice between a pre-concrete vs. post-concrete method will depend on the intended use of the data. Table 5-1 shows some factors that need to be considered before making the choice between the two methods and when each is more appropriate. The information in Table 5-1 was compiled based on the literature review of post-concrete methods presented in Section 5.2.1 below as well as the authors' experience with 3D imaging techniques that may be used as pre-concrete rebar mapping methods.

Table 5-1 Factors that affect the choice of pre- vs. post-concrete rebar mapping methods.

Factor	Pre-Concrete Method	Post-Concrete Method
Rebar area – the extent of the area to be mapped.	Large areas (tens or hundreds of meters)	Small areas (a few square meters)
Rebar depth – the maximum depth of the rebar from the concrete surface that needs to be mapped.	Deep (> 0.3 m) & shallow (< 0.3 m)	Shallow (< 0.3 m)
Rebar congestion – the amount of rebar and other objects inside the concrete that needs to be mapped.	High & low congestion	Low congestion
Level of detail – the level of detail required (e.g., do you need to know where the drains are?)	High & low levels of detail	Low level of detail
Rebar diameter – the minimum diameter of the rebar that needs to be mapped.	Small (< 20 mm) & large (> 20 mm) diameters	Large diameter (> 20 mm)
Interruption to construction schedule	Yes	No
Accuracy	Unknown	Unknown
Data acquisition time – the onsite time available to acquire the data.	Unknown	Unknown

5.2.1 Literature Review of Post-Concrete Rebar Mapping Methods

There are commercial instruments that may be used to locate rebar in concrete (IAEA-TCS-26, 2005). These instruments can use radar (e.g., ground penetrating radar or GPR), magnetic field or eddy current, ultrasound and thermography techniques, or a combination of more than one . A literature search was conducted in order to identify such technologies and to assess their suitability for the rebar mapping scenario.

Of the available technologies, GPR and magnetic field or eddy current techniques are the most common and are discussed in this section. These two methods have been used to identify structural reinforcing components in existing concrete, such as rebar and pre-stressed and post-tensioned tendons. Although these two methods were not evaluated in the testbed, brief descriptions of these techniques are included here as they are potential alternate methods for detecting rebar in hardened concrete.

Commercial rebar locators generate a primary electromagnetic field that causes electric eddy currents in a conductive material. These currents cause a secondary electromagnetic field that is then sensed by the locator's receiver. This method is limited by the ability of the target to generate an electric current and the proximity of the target to the instrument (http://www.geophysical.com/whatisem.htm). The locator must be directly over the area being scanned and as close as possible to the target to allow the primary magnetic field to penetrate as deep as possible into the subsurface. The locators are good for measuring the first layer of rebar in concrete however the electromagnetic field generated in deeper layers can be masked by shallow layers.

Some models, also called covermeters or pachometers, can measure the amount of cover of concrete over the rebar. Different models have a different maximum depth to detect metal. A locator can sense large diameter rebar with up to 0.25 m (10 in) of concrete cover and small diameter rebar with 0.15 m (6 in) of cover. For example, a #7 (22.2 mm [7/8 in] diameter) bridge deck rebar could be identified with up to 0.18 m (7 in) of concrete cover (http://www.ndtjames.com/Rebarscope-Complete-System-p/r-c-4.htm). Commercial rebar locators are relatively cheap, easy to use, and provide quick readings. Locators that estimate the amount of cover cost less than $1000, and locators that estimate cover and bar size cost between $1000 and $4000 (Riley, 2003).

GPR uses an antenna to emit high frequency electromagnetic waves that are reflected back from subsurface boundaries and received by a second antenna. The depths to the subsurface boundaries are then calculated based on the time for the emitted waves to return. The electrical conductivity of the ground affects the depth at which boundaries can be identified. In moist or clay soils, the waves dissipate quickly and depths of only a few centimeters may be achieved. Depths on the order of 9.1 m (30 ft) can be achieved in dry soil, granite, and limestone. Higher antenna frequencies result in shallow signal penetration but better subsurface resolution. For example, a 2600 MHz antenna frequency provides a depth range up to 0.3 m (1 ft) in concrete. A 1000 MHz frequency extends the range to 0.6 m (2 ft) (http://www.geophysical.com/whatisgpr.htm; Takayama, 2009).

There are four main uses of GPR for bridges and structures. GPR is used for concrete inspection to identify depth and location of metallic objects like rebar and pre-stressing and post-tensioning tendons for bridges, walls, and floors. GPR can determine the thickness of concrete structures and bridge decks. It can determine if the specified amount of concrete cover over rebar in bridge decks is in place. It can map deteriorating areas and voids for condition assessments of bridge decks and structures.

The antenna needs to be moved across, and held very close, or in contact with, the material being analyzed. This requires the area to be accessible to an operator and the equipment. This technique can identify metal and polyvinyl chloride (PVC) conduits. Metal completely reflects the signal so a fine metal reinforcing mesh will mask boundaries below it (http://www.geophysical.com/whatisgpr.htm). Shallow rebar can shadow deeper rebar depending on the antenna frequency and rebar spacing. A control unit with software is required to post-process the signal for simplified viewing and for interpretation by the operator. GPR units can be expensive and require training and experience for effective usage. Hand-held units can cost approximately $3 000 to $5 000. A walk-behind unit on wheels costs on the order of $50 000.

Both rebar locators and GPR are currently used to identify reinforcing components in construction. Both tools could be applied for comparison purposes in the IACJS Testbed as examples of existing techniques to identify rebar after pouring concrete. In the specific testbed scenario for the railway embedments, these tools could be considered as options to eliminate the need to install and remove the wood dowels which mark the target locations for the stirrups. However, both GPR and rebar locators have limitations that may prevent their use in this scenario. Neither technique is expected to work well for identifying individual bars, at different depths, in a dense field of rebar. Rebar locators tend to be more effective at identifying rebar in building floors or walls where the bars are typically spaced further apart. Although GPR with a high frequency antenna could provide sufficient resolution to identify specific rebar in the top layer, it may not provide sufficient depth resolution to indicate that a plinth stirrup could be safely drilled to the required depth.

5.2.2   Literature Review of Pre-Concrete Rebar Mapping Methods

Without a tool to efficiently and accurately identify the location for the plinth stirrups between the rebars after the concrete is poured, other technologies were required to satisfy this need. Potential technologies were 3D imaging[4] techniques such as laser scanning and photogrammetry.

---

[4] 3D imaging refers to the practice of using 3D imaging systems to measure or to capture existing conditions in the built or natural environment. 3D imaging systems are instruments that are used to rapidly obtain 3D information (e.g., 3D coordinates) about an object or a region of interest. The most common type of 3D imaging systems currently used in construction are time-of-flight laser scanners.

The following sections present a review of laser scanning (Section 5.2.2.1) and image-based 3D reconstruction or photogrammetry (Section 5.2.2.2) technologies for construction applications and in particular for the rebar mapping scenario.

5.2.2.1 *Laser scanning*

The use of laser scanning to locate rebar prior to pouring concrete does not appear to be common practice or it has not been widely documented. Besides Whitfield (2010), no published research was identified during the literature review. However, the use of laser scanning in construction for other applications has been documented extensively. Therefore, the authors of this report decided to concentrate the literature review on the use of laser scanning in modular construction, which is most closely related to the rebar mapping application.

Modular construction, which includes prefabrication, preassembly, and modularization, has been identified as an activity that could lead to improvements in construction efficiency (National Research Council, 2009; Greaves, 2006; Presley and Weber, 2009). Modular construction requires:

1. verification of the component at the factory or off-site plant to ensure that it is built to specifications and
2. determination of the site conditions
   a. to verify that they will fit to other prefabricated or in-situ components
   b. to determine potential clashes when placing the component

Bosché developed an algorithm to automatically calculate the as-built *pose* and *shape* of certain construction objects (Bosche, 2010). This approach enables dimensional checks such as verticality of columns and distances between columns. However, Bosché recommended further experiments using a more accurate instrument, shorter scanning ranges (40 m to 50 m), and comparisons with ground truth in order to draw conclusions about the accuracy of his approach.

Cianbro (Goucher, 2010) used laser scanning to construct modules in a plant in Maine that were then shipped and installed in Texas. Laser scanning was used to verify every piece of horizontal steel for elevation and all vertical columns for location and elevation. For structural steel signoff, the horizontal framing was checked in two to three places on the horizontal member for elevation, 6.3 mm (1/4 in) tolerance from design height, so that the piping would fit. Laser scanning was also successfully used to help locate where the piping [3.2 mm (1/8 in) tolerance] would go. The above tolerances would indicate that laser scanning is a suitable technology for rebar mapping since the majority of rebar on the railway bridge decks in the selected scenario have larger diameters.

Other studies on the performance of laser scanners are relevant as they provide information on the uncertainties associated with these types of instruments (Hiremagalur, 2007; Slattery and Slattery, 2010; Salo et al., 2008; Voegtle et al., 2008; Lichti, 2006; Boehler et al., 2003). Of note was the study by Sternberg and Kersten (2007). They investigated the combined uncertainty from the instrument, the registration, and the modeling on the end product. Four laser scanners were evaluated and the study concluded that the uncertainty of a resulting 3D CAD model derived from the scanned data was about 1 cm. This would indicate that these types of systems would work well for rebar with diameters of 1 cm or larger.

Clash detection (or interference checking) is another common application for laser scanning in construction (Greaves and Hohner, 2009; Longstreet, 2010; Dobers et al., 2004) and has become a widely available commercial capability. Clash detection and the rebar mapping scenario are very similar in nature since in both cases the final desired outcome is a "clash-free" path for the construction component or tool of interest.

5.2.2.2 *Image-Based 3D Reconstruction or Photogrammetry*

Image-based 3D reconstruction or photogrammetry has been applied to various construction scenarios and recently at a more accelerated pace due to advances in digital photography, the low cost of memory, and the increasing network bandwidths. This has significantly changed the way daily construction photo collections are used in the Architecture, Engineering, Construction and Facility Management (AEC/FM) community. Today, an image taken on a construction site can be quickly shared online or transmitted through email and potentially be observed by all key project participants. For example, on a typical building construction project with 15 work packages, an average of 150 to 200 photos is collected on a daily basis. Figure 5-3 illustrates several images that were collected by the authors in order to document the rebar locations prior to concrete placement on a railway infrastructure project. As seen in Figure 5-3, the images also document the locations of pipes, underlying conduits, and a manhole.

Figure 5-3 Unordered photos collected prior to placement of concrete on an actual project.

Such a large and diverse set of imagery enables the actual site to be fully observed from every possible viewing position and angle during the construction of a project. The availability of such rich imagery - which at minimal cost captures the finished or in-progress work - enables geometrical reconstruction and visualization of the actual construction models and can have broader impacts for the AEC/FM community. Over the past few years, using these emerging and already available sources of as-built documentation several researchers have created and developed image-based 3D modeling techniques which take image collections and automatically reconstruct 3D point clouds. The goal of these multi-view 3D reconstruction techniques is to infer the geometrical structure of the scene.

These techniques fall mainly into four categories:

1. *Conventional photogrammetry*: Conventional photogrammetric techniques use high-resolution analog cameras and provide accurate models comparable to those of laser scanning systems, but their spatial point density is more limited (Furukawa and Ponce, 2006) and they may require application of markers (Uffenkamp, 1993). Examples of such issues are restriction in camera rotations, range of the focal length, and analysis of the orientation data. These analog instruments have many limitations that may not affect their application to mapping applications (Moore, 1992), but restricts their utility for non-topographic applications such as construction monitoring, which requires frequent observations.

2. *Interactive 3D reconstruction techniques:* Over the past few years, several researchers and companies have developed techniques that allow users to interactively generate 3D models from digital images. In some of these techniques (e.g., Devebec et al., 1996; PhotoModeler, 2011) the user needs to manually calibrate the camera (calibration is the process which enables camera parameters such as focal length and distortion to be known), select a set of points on the scene from one of the images, and match these points to the same points observed from other images. Once several tracks of matching points are generated, through a bundle adjustment algorithm (Triggs et al., 1999), a sparse point cloud is generated and the cameras are geo-registered. These techniques also require supervision and the density of the generated point clouds may not be adequate for most engineering applications.

3. *Calibrated multi-view 3D reconstruction techniques*: In calibrated 3D reconstruction techniques, the camera position and internal parameters are assumed to be known and they are estimated directly from the image collections. By using multiple images, 3D point information can be recovered by solving a pixel-wise correspondence problem. Furukawa and Ponce (2010) describe one of these techniques that when compared to other multi-view techniques, generates 3D point clouds with higher accuracy and completeness. Similar to the previous techniques, this technique also requires manual calibration, which may impact its applicability for frequent inspections.

4. *Uncalibrated multi-view 3D reconstruction techniques*: Due to recent developments in automated feature detection and matching techniques such as Scale Invariant Feature Transforms (SIFT) (Lowe, 2004) and based on structure from motion (SfM) techniques, several research groups have developed systems which aim to automatically recover camera calibration information and reconstruct a sparse 3D representation of the scene geometry. An example of such techniques is the Phototour algorithm of Snavely et al. (2006), which is the backbone of the Microsoft Photosynth system. The SfM techniques not only reconstruct a sparse representation of the scene, but also calibrate cameras up to an unknown scaling factor. Using calibration information derived from the SfM algorithm, several techniques such as those described by Furukawa et al. (2009) and Golparvar-Fard et al. (2009) and Golparvar-Fard et al. (2010) are proposed which generate dense point clouds. Given the availability of unordered photo collections on construction sites, and the need for fully automated systems, uncalibrated multi-view 3D reconstruction techniques have the potential to be the most useful approach for AEC/FM applications.

5.2.3   Selected Technologies

The technologies that were applied toward the rebar mapping scenario and whose performance were characterized in the IACJS Testbed were laser scanning and the D4AR (4 dimensional augmented reality) system of Golparvar-Fard et al. (2009; 2010).

5.3   Description of the Experiments

The experiment that was demonstrated on June 30, 2011 involved the use of two different types of sensing techniques to map the location of rebar within a mockup of the type of rebar cage found on a railway bridge deck. The data from the two sensing techniques were processed and compared with ground truth and the results were compared and visualized using several different methods.

5.3.1   Design and Fabrication of the Reconfigurable Rebar Cage

The reconfigurable rebar cage (R2C) is a mockup of the type of rebar cage that may be found in a railway bridge deck. The R2C was designed and fabricated to easily simulate various rebar configurations. The rebar is held rigidly within the R2C using clamps in order to establish a reliable ground truth. The R2C accommodates rebar sizes from #4 to #8 with any size combination between layers with a minimum spacing increment of 7.6 cm (3 in) on center. The overall footprint is 2.44 m x 3.66 m (8 ft x 12 ft) with a usable grid area of 1.91 m x 3.2 m (6.25 ft x 10.5 ft). The rebar can be configured in one or two layers. The bottom layer is set approximately 10.2 cm (4 in) from the deck and the top layer can be reconfigured to be from approximately 15.2 cm (6 in) to 44.5 cm (17.5 in) off the deck (depending on the diameter of the rebar used). Figure 5-4 shows pictures of the conceptual model as well as the as-built version of the R2C.

Figure 5-4 The R2C 3D CAD model and as-built product.

The as-built configuration of the R2C uses epoxy-coated, #6 rebar (1.9 cm diameter). Two layers of rebar are separated by approximately 30.5 cm (12 in) and each layer consists of thirteen 3.66 m (12 ft) long rebar laid on top of twenty-two 2.44 m (8 ft) long rebar. The rebar are spaced at

15.2 cm (6 in) apart within each layer and the bottom layer of rebar is also separated from the rebar cage's floor (the plywood) by 15.2 cm (6 in).

5.3.2 Ground Truth

5.3.2.1 Point Cloud and Rebar Intersections

The rebar cage was scanned with the CLR instrument (described in Section 5.1). The point spacing used to scan the rebar cage was 3 mm. Due to time constraints, the rebar cage was only scanned from one location. Monuments, placed around the rebar cage, were measured with this instrument and the iGPS system (see Figure 2-5b and Section 5.3.2.2), and they were used to locate or transform the instrument's coordinate frame to the iGPS's coordinate frame. Once the instrument was located in the iGPS coordinate frame, all the measurements were in the iGPS coordinate frame.

The point cloud produced by this instrument was used to determine the ground truth coordinates of the rebar intersections. This was a manual process and was done for each rebar intersection individually. The process involved fitting cylinders, where the radii of the cylinders were not fixed, to the point cloud of the rebars. For a given cell, only the points around that given cell were used to fit the cylinders (see Figure 5-5). The center lines of the cylinders were then projected onto a plane parallel to the top rebar layer, and the intersection of these lines was determined. The intersection points were then translated to the base plane (plane fitted to the floor points). All of these points are used to represent the ground truth coordinates of the rebar intersections in iGPS coordinates.

Figure 5-5 Picture showing a section of the point cloud of the top rebar. The points in white were used to fit the cylinders to determine the coordinates of the rebar intersections.

5.3.2.2 iGPS

The iGPS sensor is an infrared laser-based positioning system that uses a series of tripod-mounted laser transmitters whose signals are detected using an optical receiver. Once the system is calibrated, the receivers can triangulate their 3D position as long as they have line-of-sight to at least two transmitters. A receiver can calculate its position up to 40 times per second. The iGPS system installed in the IACJS Testbed has a 3D position uncertainty of ± 0.250 mm and a maximum range between a receiver/transmitter pair of 40 m.

The iGPS system was used to tie all of the measurements (including the ground truth measurements) into a common coordinate system (see Section 5.3.2.4).

5.3.2.3 Registration

As mentioned in Section 5.3.2.2 above, all of the measurements during the experiments were tied to a common reference frame (see below) using data from the iGPS for registration. This process involved measuring targets with the technology being evaluated and then measuring these same targets with the iGPS instrument. The type of targets used varied for the different technologies and are described in Section 5.3.3 for laser scanning and 5.3.4 for photogrammetry.

5.3.2.4 IACJS Reference Frame

Measurements within the IACJS Testbed are all conducted within a common coordinate system (the IACJS Testbed Reference Frame or the iGPS coordinate system). In order to establish this reference frame, eight monuments (reference points) were established in the testbed. Each monument is defined as the center of a 12.7 mm (0.5 in) sphere placed inside a kinematic nest on the testbed's walls or floor (Figure 5-6). The nominal locations of the monuments in the testbed are depicted in Figure 5-7.

(a)   (b)

Figure 5-6 Pictures of two IACJS Testbed monuments on the wall (a) and on the floor (b).

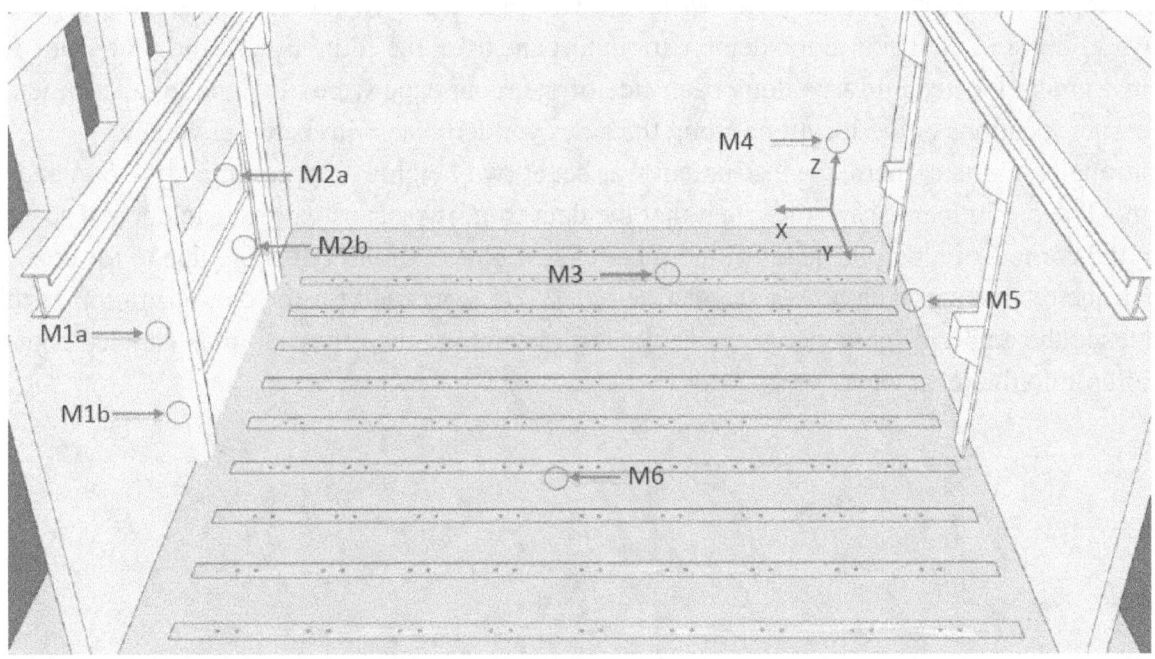

Figure 5-7 Nominal locations of the IACJS tesbed monuments and coordinate system.

In order to establish the reference frame, the locations of the monuments were measured using the iGPS system. The origin of the IACJS Testbed reference frame was then defined as the projection of monument M4 (in Figure 5-7) onto the ground plane (a plane fit to a series of 20 points measured throughout the testbed's floor with the iGPS system) in the direction parallel to the ground plane's normal. The direction of the reference frame's Z-axis was defined as this plane's normal direction. The direction of the reference frame's Y-axis was defined as the vector perpendicular to the Z-axis and along the direction from monument M4 to monument M5. Finally, the direction of the X-axis was taken as the cross product of the Y and Z axes.

Because the iGPS requires periodic recalibration, the eight monuments are re-measured after each recalibration in order to ensure that any subsequent iGPS measurements remain in the same IACJS Testbed reference frame. This assumes that the positions of the monuments do not move over time more than ± 0.250 mm (the uncertainty of the iGPS system). A preliminary analysis of the position of the monuments over time (two measurements of the monument positions were done using the coherent laser radar separated by a period of around four weeks) found that any movements were below ± 0.250 mm.

5.3.3   Data Collection from a Laser Scanner

The laser scanner used to measure the rebar cage is a time-of-flight, phase-based 3D imaging system with a maximum range of 79 m. It has a manufacturer specified measurement position uncertainty of ± 5 mm for ranges less than or equal to 25 m; the range at which the rebar cage was scanned was less than 7 m. The instrument was placed at two corners of the rebar cage as shown in Figure 5-8. Prior to the demonstration, scans from the four corners and four scans with the instrument located midway along each side of the rebar cage were obtained to determine if the corner locations or the locations along the sides yielded data with better coverage. Additionally, at the corners, the instrument was set at two heights – approximately 1.5 m and 2.2 m. The preliminary scans indicated that the data from two scans were sufficient and scans from the corners or along the sides of the rebar cage yielded similar coverage due to the small size of the rebar cage. The higher scanner height, 2.2 m, was used for the demonstration as this decreased the angle-of-incidence between the laser beam and the rebar and thus enabled better visibility into the rebar cage.

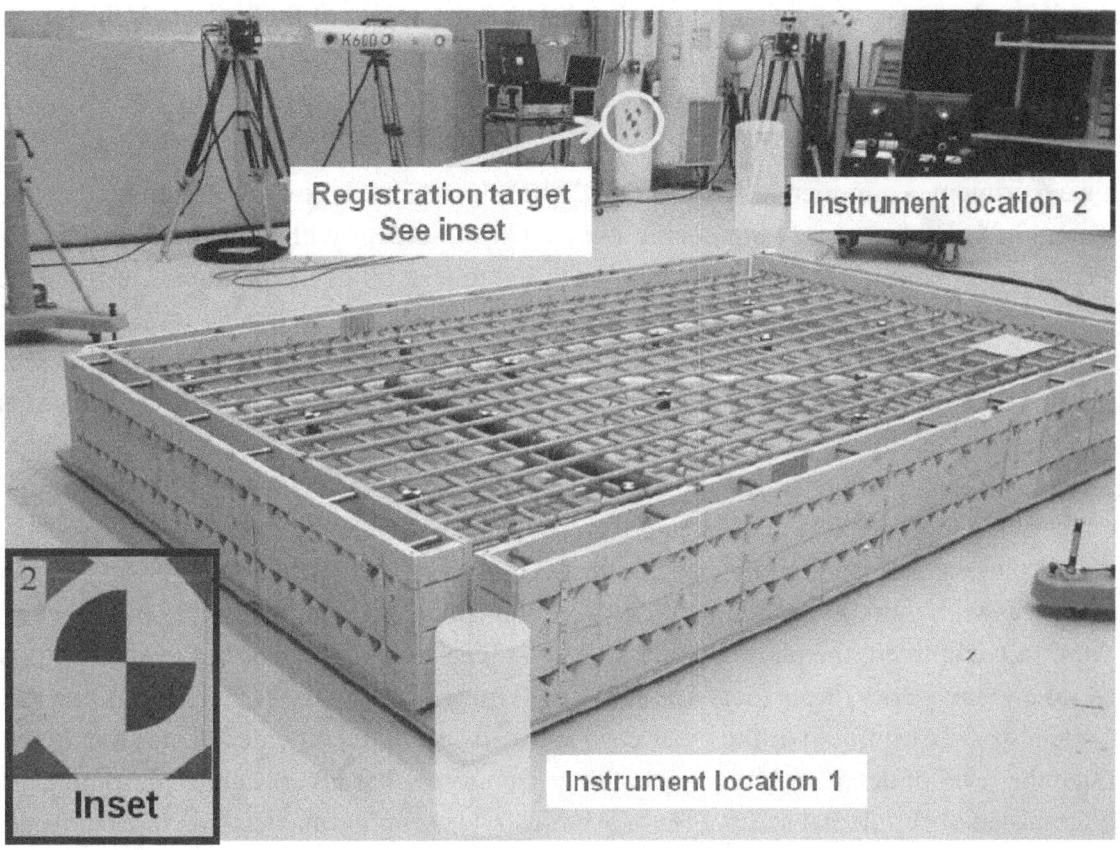

Figure 5-8 Photo showing the two locations of the laser scanner relative to the rebar cage. The photo also shows one of the targets used to register the two scans to each other and to the iGPS coordinate frame.

The procedure to scan the rebar cage consisted of leveling the laser scanner on a metrology stand, obtaining a preview scan (point spacing about 50 mm at 10 m) of the entire area, defining a region which contains just the rebar cage, scanning that region with the point spacing set at 6 mm at 10 m, and scanning eight targets (one such target is shown in Figure 5-8) that were placed in various locations around the rebar cage. Each target was identified as a target and was given a unique name or target ID. After each target was scanned, the software (provided by the manufacturer and which also controls the instrument) automatically executed an algorithm to extract a single point representing the center of the target. The instrument was then moved to the second location, and the procedure was repeated.

The eight targets were also measured with the iGPS system using a handheld probe with a point tip (see Figure 2-5b). In this case, the tip was placed at the center of the target and the target center was measured directly, that is, the center point was not derived.

5.3.4 D4AR Image-based 3D Reconstruction Setup

For this experiment, a commercially available digital single lens reflex (SLR) camera was used. The camera has a 36.0 mm x 24.0 mm single-plate CMOS sensor with approximately 21.1 million effective pixels. A 20 mm fixed lens was used to take all of the pictures. During the data collection, the ISO speed of the camera was set to 600 to ensure a reasonable level of brightness across all images. Over 850 wide-range and close-up images were taken to minimize the visual occlusion of the bottom layer of the rebar while maintaining a normal distribution of the images across the rebar cage. In order to reduce the computational time for the processing step, 380 images were randomly selected. During the experiment, the camera was installed on a frame which had two iGPS receivers to track and record the location and the orientation for each image. The purpose of installing the iGPS unit was to compare the actual location and orientation of the cameras with those generated by the D4AR algorithm. To ensure all areas of the rebar cage were fully captured, the photographer slowly walked around the rebar cage and captured an average of one image per linear foot. About 50 to 70 images were also captured from the top to ensure the areas in the middle of the rebar cage were also fully documented. This strategy helped maximize the level of detail and fully document the bottom rebar layer. Table 5-2 summarizes the image-based experimental setup. Figure 5-9 illustrates several images that were captured of the rebar cage in the IACJS Testbed.

Table 5-2 Camera setup for the D4AR image-based 3D reconstruction.

Setting	Description
Camera Type	Digital SLR
Lens	20 mm – f/2.8
ISO speed	ISO 600
Images	3 images/linear meter around the rebar cage plus 50 to 70 images from the top to capture the middle of the rebar cage
Spatial resolution	Approx. 21.1 megapixels
Total # of images	850

Figure 5-9 Several images of the rebar cage collected in the IACJS Testbed.

A total of fifteen rebar targets were used to transform the outcome of the image-based 3D reconstruction into the site coordinate system. Figure 5-10 and Figure 5-11a illustrate the iGPS receiver unit along with the camera, and Figure 5-11b shows three of the rebar targets.

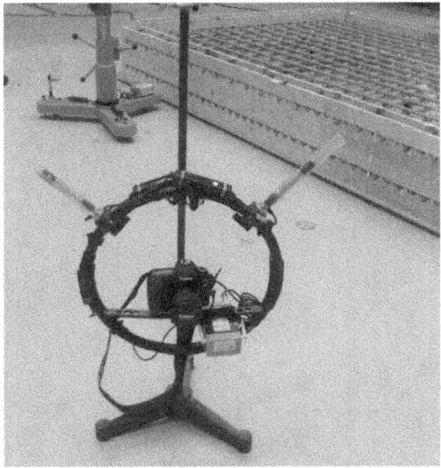

Figure 5-10 The camera with the iGPS receiver frame.

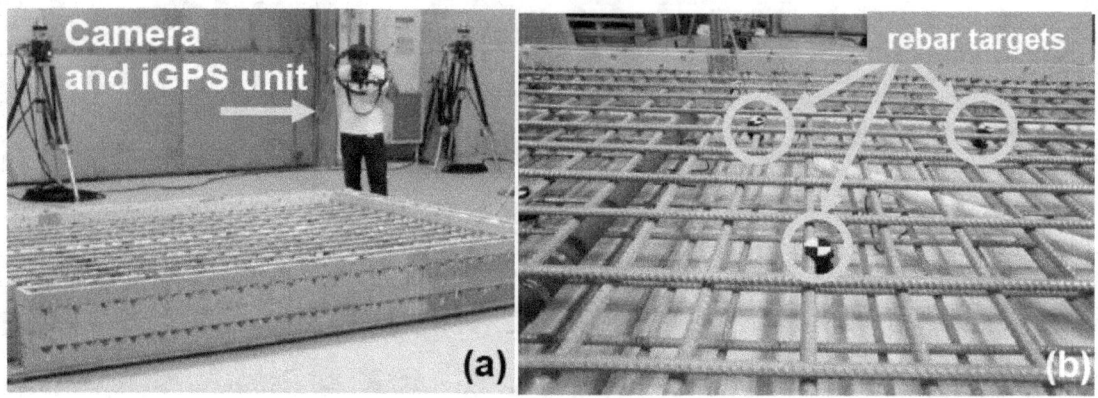

Figure 5-11(a) The camera and iGPS unit in use; (b) the rebar targets.

5.3.5   Drill

The drill in the demonstration was fitted with a pair of iGPS receivers, as shown in Figure 5-12, which were tracked by the iGPS system. The iGPS receivers were mounted such that they were oriented along the longitudinal axis of the drill bit and were connected to a shoulder/waist strap, which also holds the computer that performs the position calculations. The receivers' positions are then wirelessly transmitted to the iGPS server. The iGPS receivers' positions relative to the drill bit's tip are known and are preprogrammed into the iGPS server, which calculates the position and orientation of the drill bit's tip. This information is used by the drill feedback application to determine whether it is safe or unsafe to drill at a particular location. The iGPS server has the ability to handle scenarios with multiple drills in order to simulate a drilling crew working simultaneously on the bridge deck.

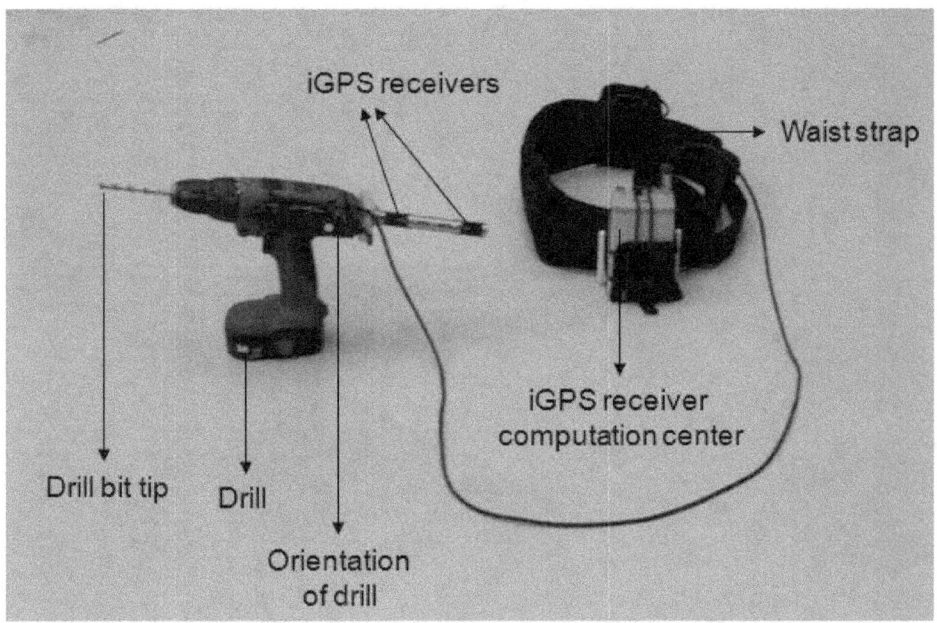

Figure 5-12 The modified drill setup with the iGPS receivers attached to the drill.

5.3.6   Rebar Targets

Fifteen targets were placed on the top rebar layer of the rebar cage. The targets consisted of a circular black and white sticker mounted on top of a round clamp as shown in Figure 5-13. The locations of the targets were selected to maximize the coverage across the top face of the R2C. The locations of the centers of the targets were measured using the iGPS system.

Figure 5-13 The rebar targets and a projected laser pattern.

5.3.7 Modeling

5.3.7.1 Rebar Model

The specific rebar cage configuration in the IACJS testbed was modeled as a CAD model as shown in Figure 5-14.

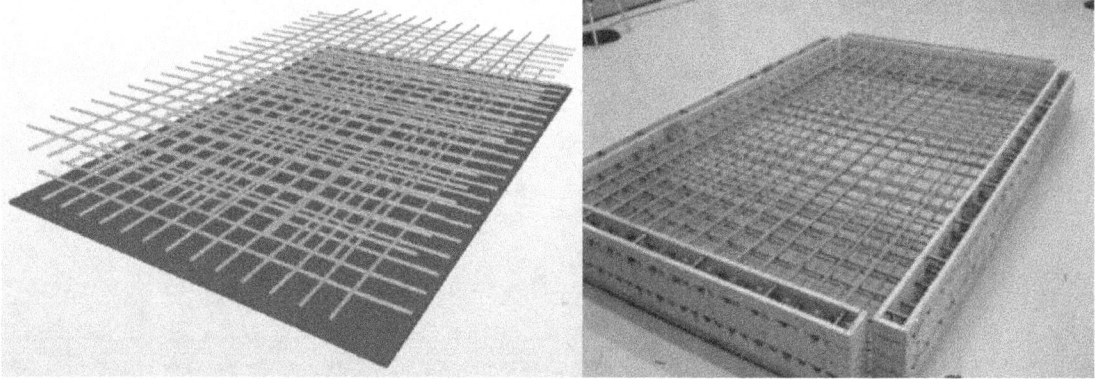

Figure 5-14 The CAD model (left) of the rebar cage configuration (right) in the IACJS Testbed.

5.3.7.2 Bridge Model

A CAD model of a bridge was created to help visualize the rebar cage within the bridge structure. The bridge was modeled as an extrusion of its cross-section. The relevant bridge as-built drawings were imported into a CAD modeling software and the cross-section of the bridge was traced to form a closed poly-line. This poly-line was extended along a curve in order to form an extruded model of the bridge. A region of space equivalent to the usable volume of the reconfigurable rebar cage was then subtracted from the bridge deck in order to model an opening in the bridge with visible rebar.

5.3.7.3 Model Registration

The bridge deck model and the rebar cage model were grouped together and registered to the iGPS coordinate system. In order to register the model to the appropriate coordinate system, the corner points of the rebar cage in the IACJS Testbed were measured using the iGPS system. The rebar cage and bridge models were then rotated and translated appropriately such that the corner points of the rebar cage model were at the same location, in the CAD modeling environment, as their corresponding corner points in the iGPS coordinate system. The registered model was then exported as an Industry Foundation Classes (IFC) file. The IFC file was then used to visualize the bridge deck with the embedded rebar cage as shown in Figure 5-15 using an IFC viewer.

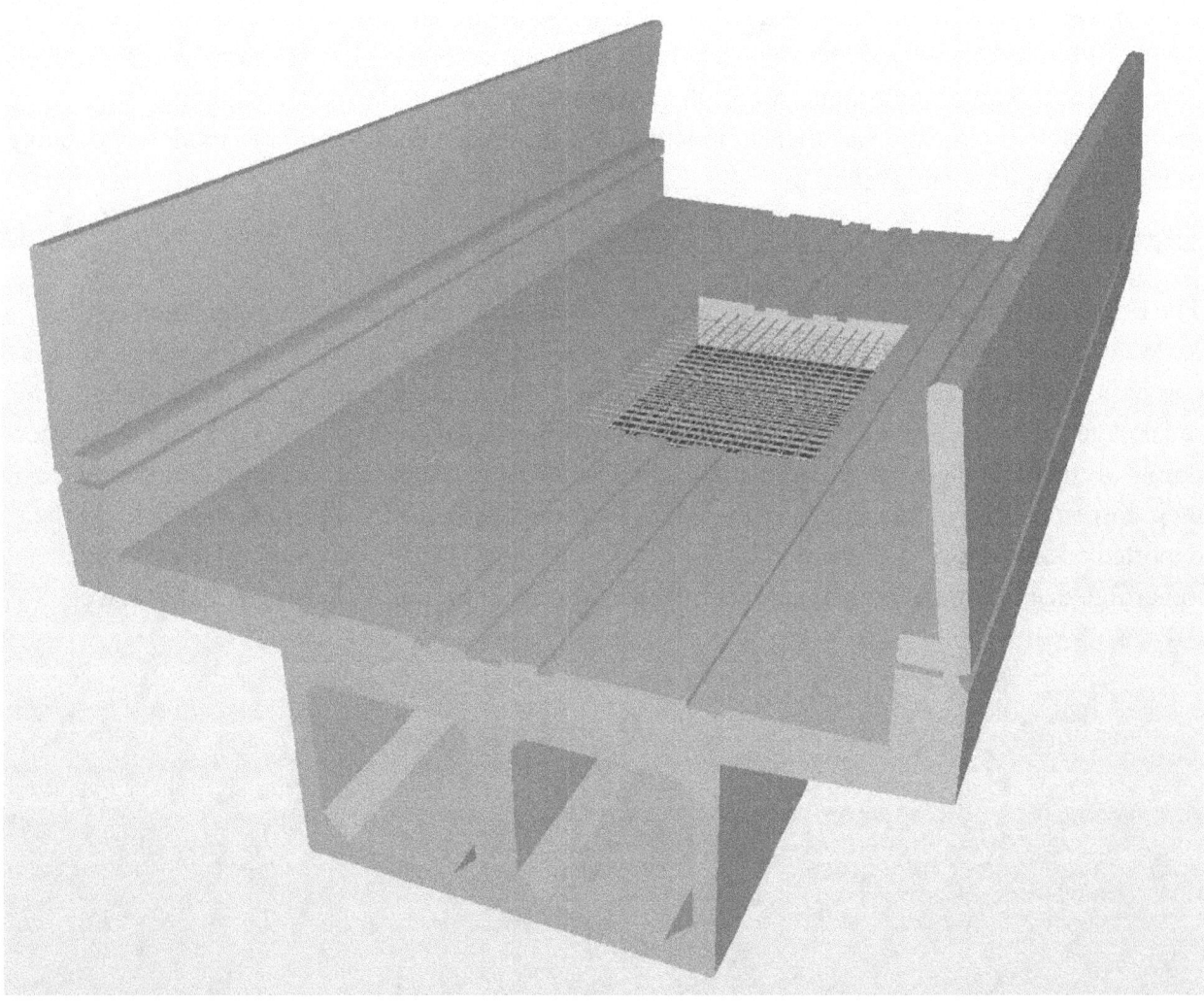

Figure 5-15 The BIM of the bridge deck with the embedded rebar cage as seen through an IFC viewer.

5.3.8 Visualization

5.3.8.1 BIM

The rebar cage and bridge models described in Section 5.3.7.2 above are visualized as a BIM by exporting an IFC file from the software that was used to generate the models. IFC is the open-standards product data model and file exchange format, developed by buildingSMART (http://www.buildingsmart.com/), that is used to facilitate interoperability between software applications in the building and construction industry. The IFC representation of the rebar and bridge models is used, rather than commercial BIM software, to simplify the visualization of the models and associated safe drilling zones.

58

The rebar and bridge models are shown in Figure 5-15. An IFC viewer was used to visualize the IFC model of the rebar cage and bridge model. The position and orientation of the IFC rebar cage model in the bridge model corresponds to the coordinate system used for the experimental scanning of the physical rebar cage. A rectangular section of the bridge deck is removed to show the rebar cage.

5.3.8.2 Laser Projector

In order to visualize the results from the experiments on the actual rebar cage in the IACJS Testbed, a laser projector (Figure 5-16) was used to project patterns onto the rebar itself. In Figure 5-13, a square pattern is projected onto a piece of paper that is lying flat directly on the rebar cage. The projected pattern in the figure is a square corresponding to the square formed by the centerlines of the four rebar lengths directly underneath. The position and orientation of the laser projector is tracked by the iGPS system and the patterns are produced using data from the actual rebar cage.

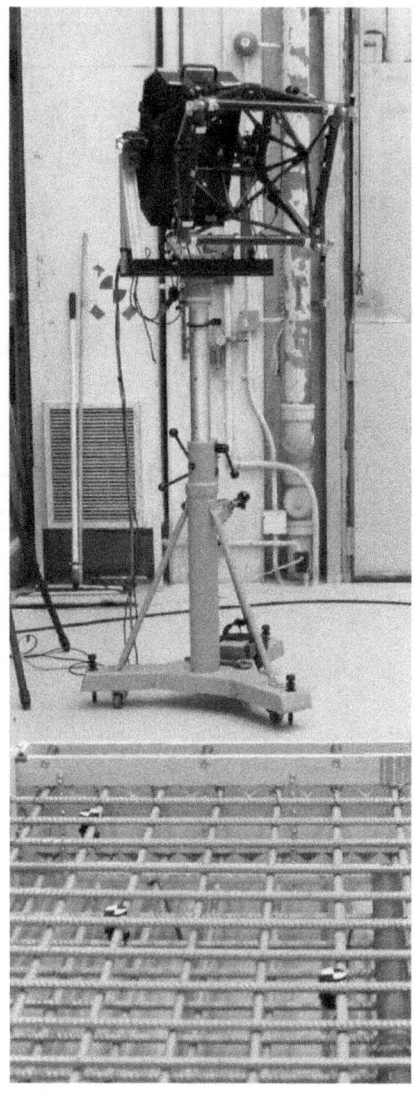

Figure 5-16 The laser projector used to visualize results on the actual rebar.

5.3.8.3 Drill Feedback

The drilling feedback application developed for the IACJS Testbed determines, in real time, whether it is safe or unsafe for a drill to continue drilling at a particular location in the concrete deck. The application uses information regarding the position and orientation of the drill bit's tip from the iGPS server and the information regarding the zones safe for drilling from the IFC data file with information regarding safe void zones as determined by the rebar mapping algorithm (see Section 6.3). The drilling feedback application interface is shown in Figure 5-17.

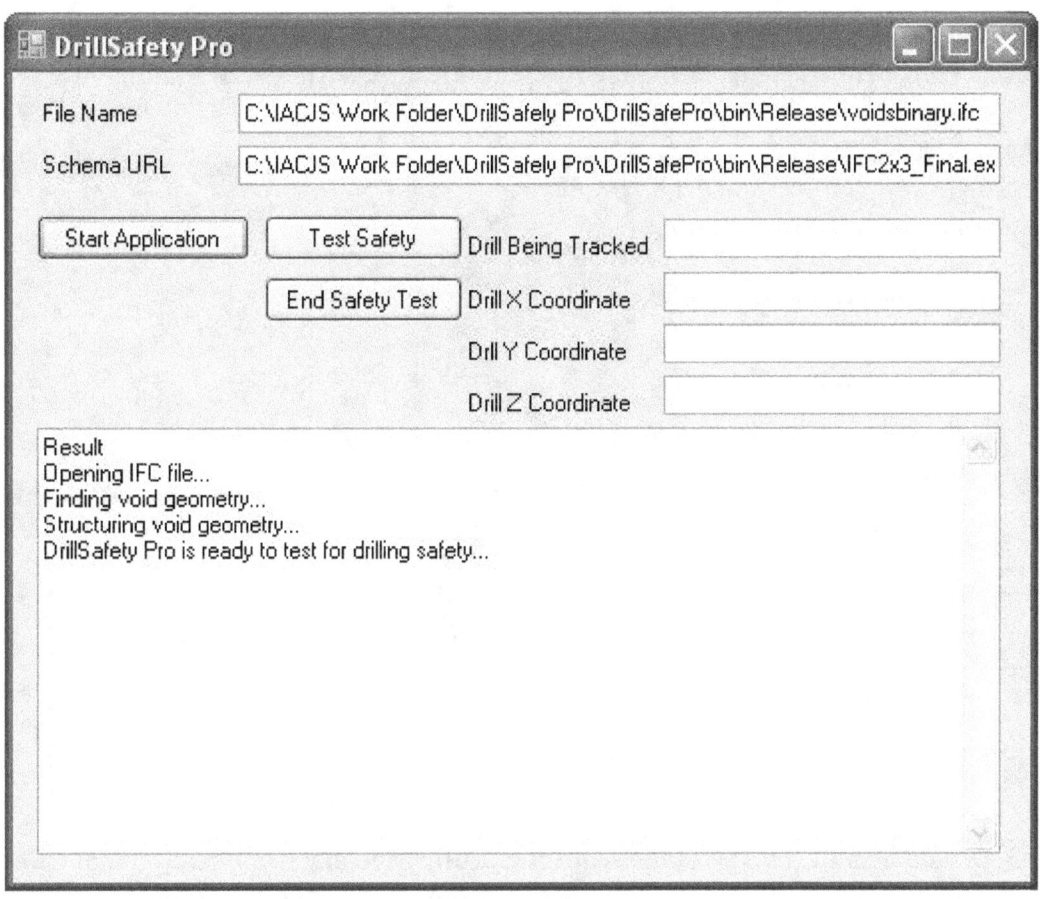

Figure 5-17 The drilling feedback application user interface.

The user inputs the 'File Name' of the IFC file which contains information regarding the zones safe for drilling and the URL of the IFC schema used in the IFC file. Upon clicking the 'Start Application' button, the application reads the IFC file and interprets the data regarding the 'safe void zones'. The application interprets each safe void zone as a generic cuboid with eight corner points. The user can then choose to test whether it is safe or unsafe to drill at the current position by clicking on the 'Test Safety' button. The application then performs collision detection between the monitored drill bit's tip and all the safe void zones to determine whether the drill bit's tip is within any of the safe void zones. The application displays the name[5] and the position (x, y, z coordinates in the iGPS coordinate frame) of the drill being monitored. When the drill is in a safe position, i.e., the position of the drill bit's tip is inside a safe void zone, the application displays a message that it is safe to drill in that location. However, when the drill bit's tip position is outside all safe void zones (for example when the drill bit's tip is over a rebar or in a

---

[5] The application can be modified to track several objects simultaneously.

zone with utility pipes), the application displays a message that it is unsafe to continue drilling in that location. The overall schema of the feedback application is shown in Figure 5-18.

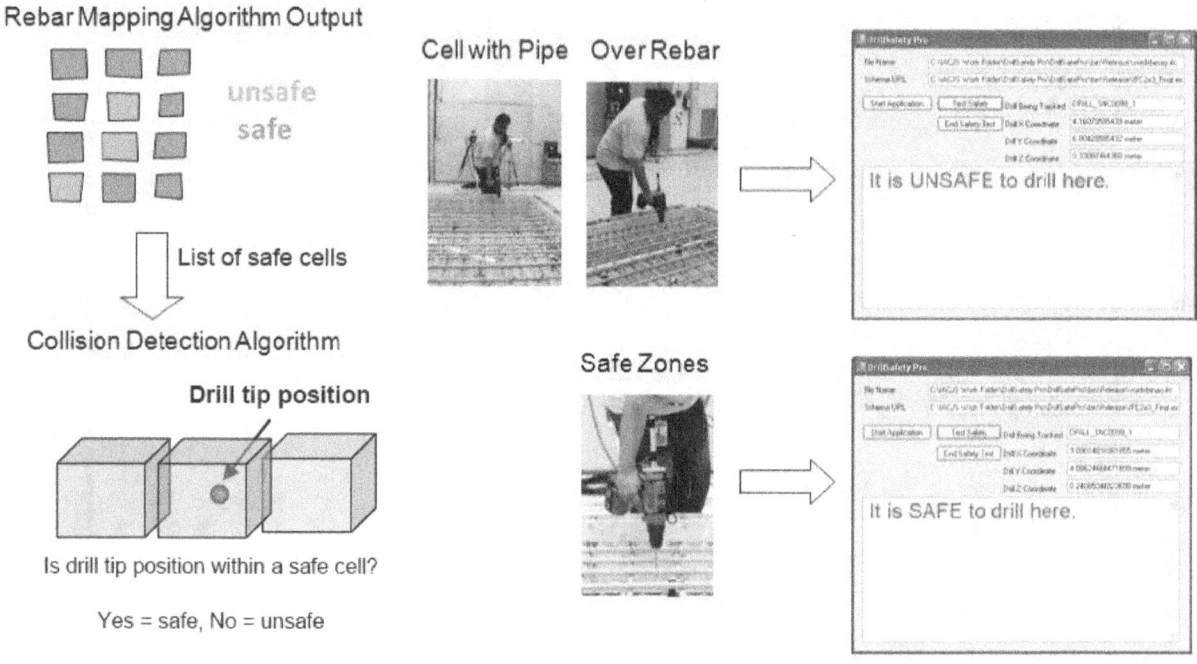

Figure 5-18 The overall schema of the real-time drilling feedback application that determines whether is it safe or unsafe to continue drilling at a location.

6 Data Processing

6.1 Laser Scanning

The only data processing done on the laser scanner data was to register the two scans and to segment the point for export. The registration process was a two-step process. The first step was to register the two scans to each other using software provided by the instrument manufacturer. This registration process was straightforward since the targets were already identified in each scan, and the registration process consisted of a few mouse clicks. The average registration error (i.e., the average distance between corresponding targets) was 3.4 mm with a standard deviation of 1.2 mm.

The second step was to get the combined point cloud into the iGPS coordinate frame. Once the two scans were registered to each other, a text file containing the target ID and the coordinates (x, y, z) of the targets in iGPS coordinates (measured as described in Section 5.3.2.2) was imported into the same point cloud software. The combined point cloud was then registered so that it was in the iGPS coordinate frame. This procedure also consisted of only a few mouse clicks. For this registration, the average registration error was 5.8 mm with a standard deviation of 2.5 mm.

The point cloud was then segmented so that only those points within the rectangle enclosing the rebar cage (Figure 6-1) were selected and these points were exported as an ASCII text file. A side view of the exported point cloud is shown in Figure 6-2. No other filtering or removal of outliers was performed.

a. Combined point cloud.   b. Selected points.   c. Segmented points for export.

Figure 6-1 Segmentation of point cloud to obtain points of rebar cage.

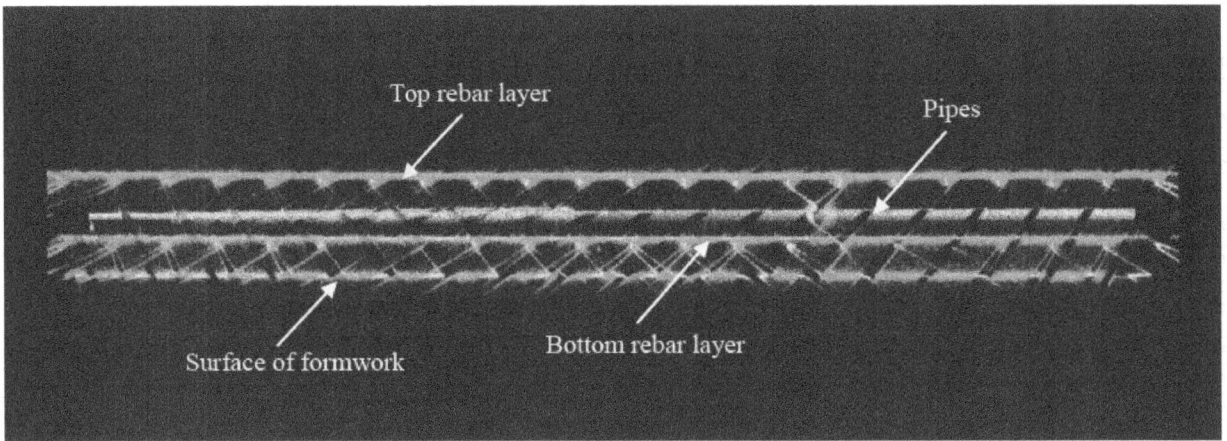

Figure 6-2 Side view of the point cloud that was exported to the algorithm to extract the rebar intersections, see Section 6.3. The points on diagonal lines are mixed pixels (or non-existent) points caused by the splitting of the laser beam as it hits the edges of the rebar.

6.2   D4AR Image-based 3D Reconstruction

As mentioned in Section 5.2.2.2, several techniques could be utilized to generate a point cloud from a set of images. In this particular study, the D4AR method (Golparvar-Fard et al., 2009; Golparvar-Fard et al., 2010) are used. The D4AR method which is mainly developed to track construction progress, first reconstructs a 3D point cloud of the scene. Next, the reconstructed scene is superimposed on a building information model (BIM) which represents the expected status of the project. The transformed point cloud is improved by a multi-view stereo algorithm. Using a voxel-coloring/voxel-labeling algorithm, different areas in the scene are labeled for actual occupancy and expected visibility. Finally a support vector machine (SVM) algorithm utilizes these measurements and labels each BIM element with a binary decision about its status (progress/no-progress). Figure 6-3 represents the data and process in the D4AR modeling and automated progress monitoring system.

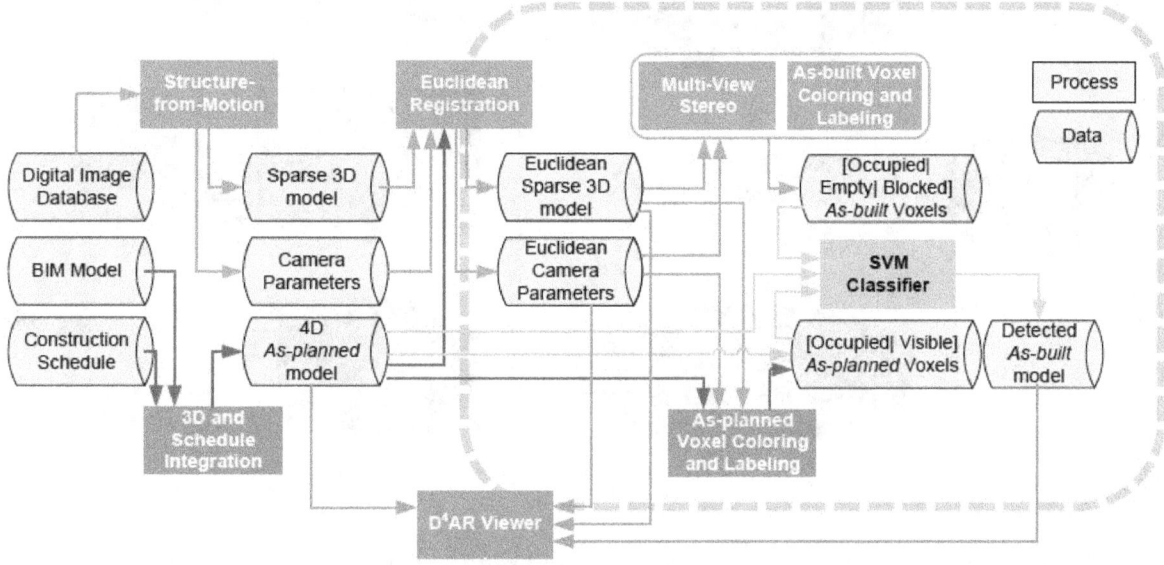

Figure 6-3 Data and process in the D4AR modeling and automated progress monitoring system (Golparvar-Fard et al., 2011).

In the IACJS Testbed demonstration, two particular components of the D4AR modeling system were used: (1) image-based 3D modeling and (2) superimposition of the reconstructed point cloud with BIM.

Figure 6-4 illustrates these modules and their immediate outcomes. In the following sections these components are described in more detail.

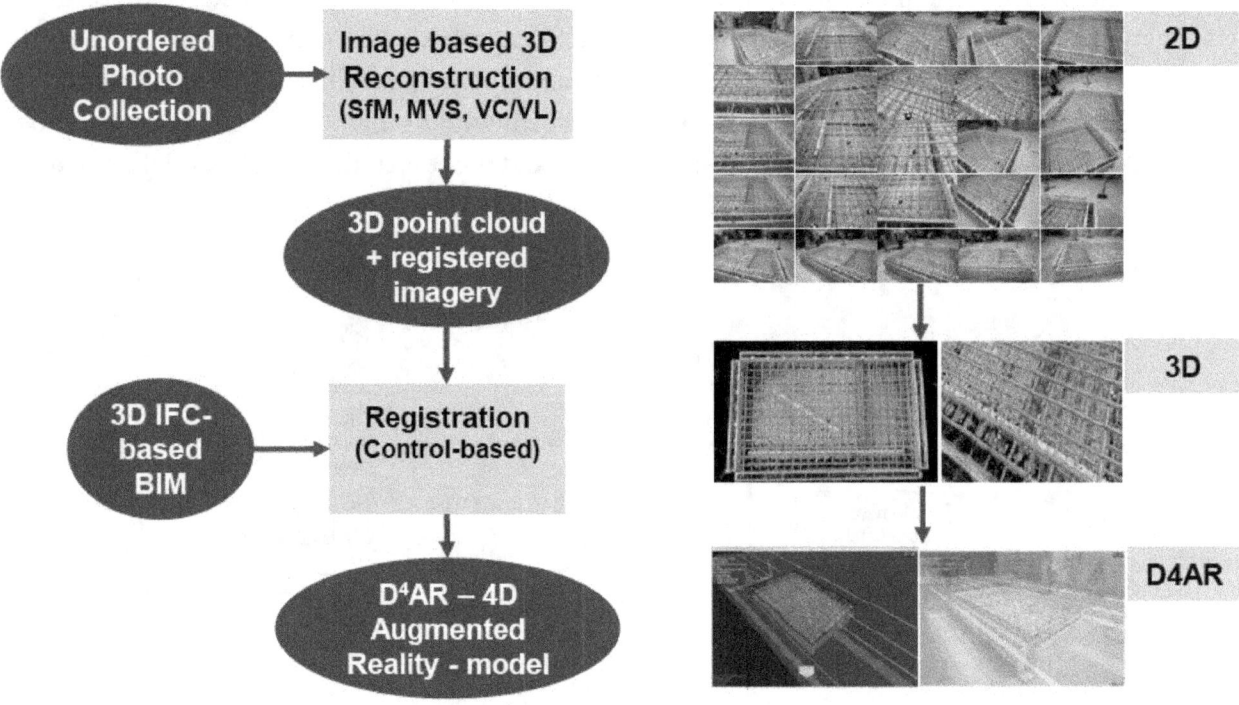

Figure 6-4 The image-based 3D reconstruction and registration modules. The outcome of each module is illustrated on the right side of this figure.

6.2.1 Generate the Image-based 3D Point Cloud

The proposed image-based modeling technique builds upon a set of Structure-from-Motion (SfM) algorithms where the objective is to reconstruct the scene without any strong prior knowledge. The developed module consists of three main components:

(1) SfM algorithm;
(2) Multi-view Stereo, and
(3) Voxel Coloring/Labeling algorithms.

The SfM module consists of the following steps:

(1) Analyzing images and extracting Scale Invariant Feature Transform (SIFT) feature points (Lowe, 2004);
(2) Matching image features across the image set (Hartley and Zisserman, 2004);

(3) Finding an initial solution for the 3D locations of these feature points, calibrating cameras for an initial image pair; and

(4) Reconstructing the rest of the observed scene while estimating the motion of the cameras based on bundle adjustment algorithms (Nistér, 2004; Triggs et al, 1999).

The outcome of the SfM module includes calibration of all cameras and generation of an up-to-scale sparse point cloud. The resulting point cloud is improved by a multi-view stereo algorithm (Furukawa and Ponce, 2010). Next, the reconstructed scene is quantized into small volumes of space (voxels). A voxel coloring/labeling algorithm marches through all voxels, analyzes each for consistent visibility, re-projects the occupied areas back on the images, and finally labels the voxels and their corresponding 2D pixels for occupancy (Golparvar-Fard et al., 2010). The outcome of this process is an up-to-scale dense point cloud.

In the IACJS Testbed demonstration, 380 images of the rebar cage were used in which the subjects were mainly the reinforcement bars and their configurations. The spatial resolution of these images was synthetically reduced to two mega pixels to test the robustness of the proposed method to low-resolution images and to minimize the computational time. The outcome of the image-based 3D reconstruction module is shown in Figure 6-5. The reconstructed point cloud consists of 9 264 880 points for the entire scene. The majority of the reconstructed points as observed in Figure 6-5a belong to the rebar cage.

Figure 6-5 The point cloud reconstructed using the image-based 3D modeling component of the D4AR modeling system.

Figure 6-6 shows a closer view of the entirety of the point cloud. The entire rebar cage and particularly both layers of the reinforcement bars, the embedded pipes, along with the form and rebar targets are all reconstructed in 3D.

Figure 6-6 The image-based rebar cage point cloud.

Figure 6-7 further illustrates the reconstructed point cloud from two synthetic views. These views were not directly photographed and the illustrations are snapshots of the 3D point cloud. As seen in this figure, all the underlying components of the rebar cage including the pipe and their seats are fully reconstructed in 3D.

(a)

Markers fully
reconstructed

Pipe

Pipe & Seat

(b)

~complete
Reconstruction with
limited line of sight

Figure 6-7 The reconstructed point cloud observed from two synthetic views.

6.2.2   Transform the Point Cloud into the Site Coordinate System

In order to transform the point cloud to the iGPS coordinate system and/or superimpose it with the BIM elements, the transformation between these two coordinate systems needs to be determined. This registration is usually represented as a rigid-body transformation. Since the SfM algorithm can only reconstruct a 3D point cloud up to an unknown uniform scale; the rigid-body transformation should include the rotation and translation as well as the uniform scaling factor (i.e., 7 DOF). Three points known in both coordinate systems is theoretically sufficient to determine these seven unknowns. However, in practice, these measurements are not always exact and using more than three points can reduce the registration error. As mentioned in Section 5.3.4, in the IACJS Testbed demonstration, 15 rebar targets were used for registration. Let n be the number of rebar targets that will be used for registration of the point cloud into the iGPS coordinate system. The points in these coordinate systems can be denoted by $\{r_{p,i}\}$ and $\{r_{iGPS,i}\}$, respectively, where i is the number of corresponding targets which ranges from 1 to n. The transformation can be written as follows [Eq. 6-1]:

$$r_{iGPS} = sR(r_p) + T \qquad \text{6-1}$$

where s is a uniform scale factor, T is the translational offset and $R(r_p)$ is the transformed version of the initially reconstructed point cloud. Minimization of the sum of the squared errors is formulated as:

$$\sum_{1}^{n}\|e_i\|^2 = \sum_{1}^{n}\|r_{i,iGPS} - sR(r_{i,p}) - T\|^2 \qquad \text{6-2}$$

To solve for this transformation, the approach proposed in Horn (1987) is used which gives a closed-form solution to the least square problem of absolute orientation. In this approach, the registration accuracy is insensitive to how the control points are selected. In the IACJS Testbed demonstration, more than three rebar targets (the minimum required number) were used to minimize the interactive selection errors. The resulting registration error was 0.4 mm.

6.3   Rebar Intersection Extraction Algorithm from Point Clouds

Once the laser scanning and the photogrammetry/D4AR technologies are applied to the rebar cage the resulting point clouds must be processed in order to extract the rebar intersections and calculate the safe drilling depth within each rebar cell. An algorithm was developed to do this automatically in order to reduce the data processing time and to ensure that the datasets from all of the tested technologies are processed in the same manner.

The input to the algorithm are:

1. the estimated radius R of the rebar
2. a segmented point cloud S
3. a unit vector w in the drilling direction (i.e., perpendicular to the drilling plane) see Figure 4-1

4. the maximum drilling depth D_{max} , see Figure 4-1

The point clouds require prior manual segmentation to retain only points inside the formwork; no other filtering or outlier removal is required. Ideally, the unit vector w is obtained by scanning the base plane prior to installing the rebars and fitting a plane to this data. If this initial scan is not possible, the unit vector w can be obtained by fitting a plane to the manually-segmented data from point cloud S.

The drilling depth is defined from the top of the drilling plane. Maximum drilling depth D_{max} is the distance between the drilling and base planes, see Figure 4-1. This distance can be obtained if both planes are scanned; the base plane before the rebar cage is placed and the drilling plane after concrete is poured. Then, by fitting a model of two parallel planes to the merged scans, D_{max} could be calculated. However, such a scenario is very labor intensive as it requires multiple scans during different construction phases. Another way would be to set D_{max} equal to the design slab thickness or wall width.

Figure 6-8 shows a flowchart of the algorithm and a brief description of each block in the figure follows.

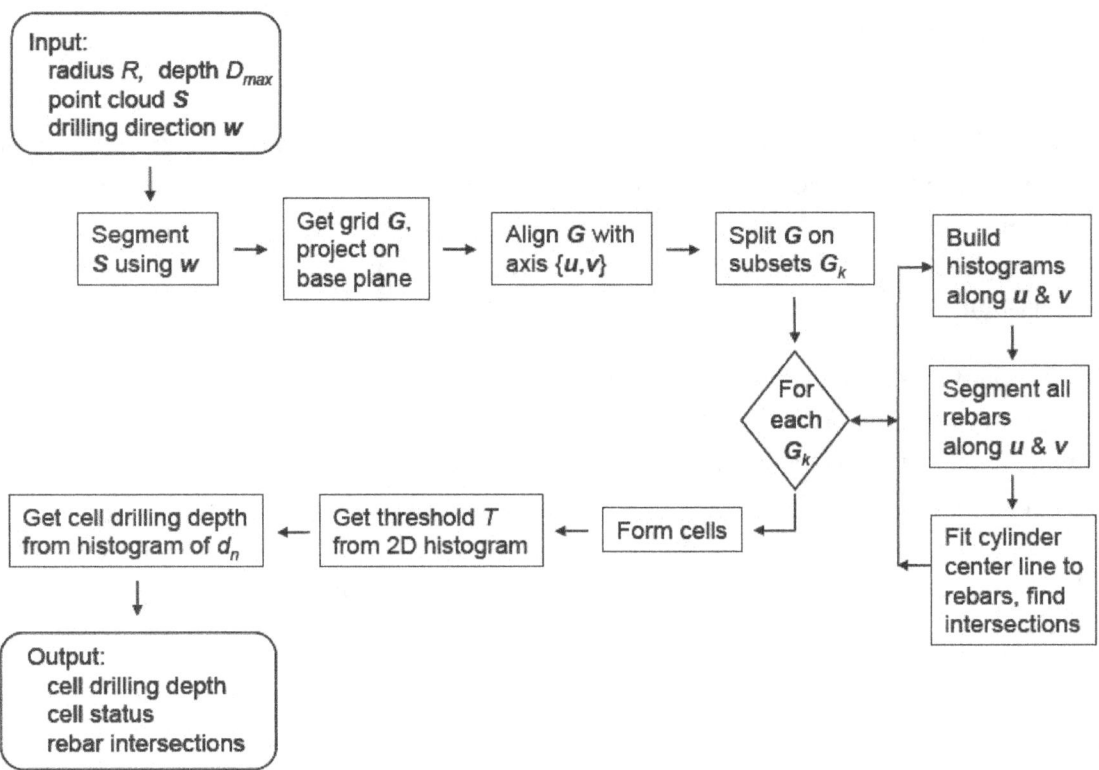

Figure 6-8 Flowchart of the algorithm for extracting rebar intersections from a point cloud and for determining safe drilling depths.

1) Segment S using w

This section of the algorithm constructs a histogram of components of the vectors in S along the drilling direction. For each vector P_n (where $n = 1,..., N$ and N is the total number of points in S) a component p_n of P_n along w is calculated, $p_n = P_n \bullet w$. Then a histogram of all p_n is built, as shown in Figure 6-9. In this figure the first peak from the left represents the points on the base plane, the second peak represents points on the bottom lower layer, the third peak represents the points on objects between the upper and lower rebar layers, and the fourth peak represents the points on the upper rebar layer. The peaks in a histogram of components p_n are used to segment the various rebar layers in S.

2) Get grid G

The centroid, B_0, of the points associated with the base plane (peak "a" in Figure 6-9) is calculated. This centroid and the unit vector w mathematically define the base plane. The points from the upper rebar layer, as segmented in Step 1, are projected along the direction w onto the base plane. These projected points on the base plane constitute grid G.

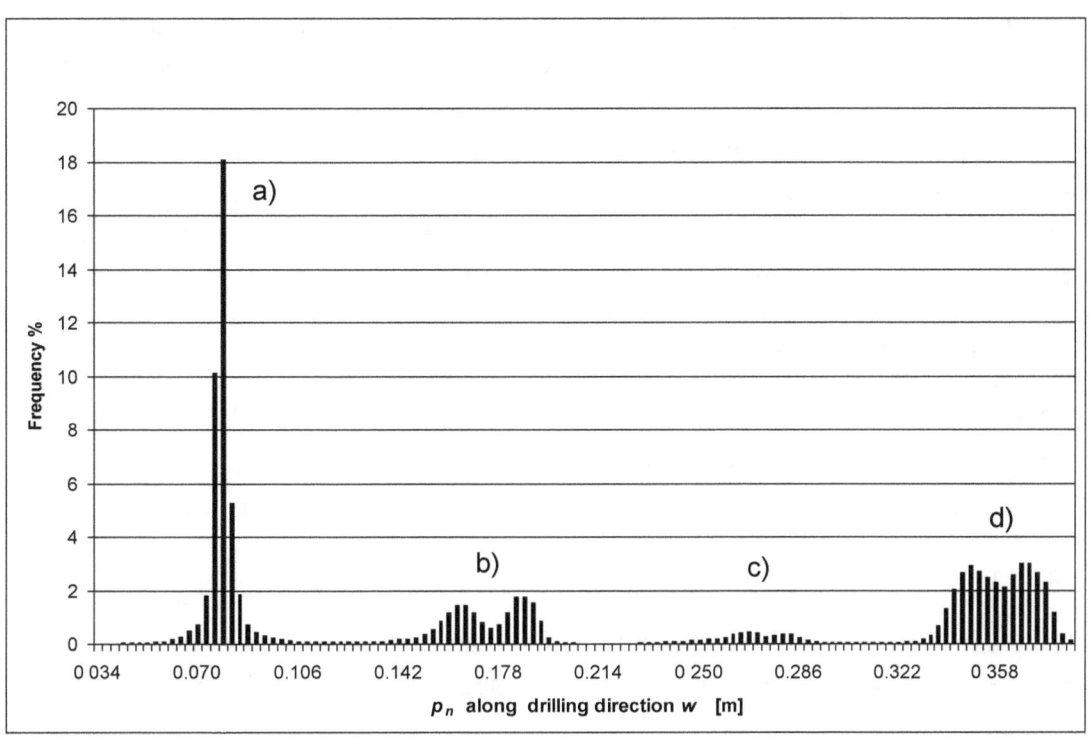

Figure 6-9 Histogram of the components of vectors P_n from point cloud S along w. The first peak (a) represents the base plane, the second peak (b) represents the lower rebar layer, the third peak (c) represents objects between the two layers of rebar (pipes in the demo), and the fourth peak (d) represents the upper rebar layer. Peaks b) and d) have two peaks which show contributions from the longitudinal and transverse rebar in the lower and upper layers, respectively. The bin size is equal to 0.3R.

3) Align G with axis $\{u,v\}$

A new coordinate system $\{u,v,w\}$ is defined so that the u and v axes align with the longitudinal and transverse axes of the rebar. Vectors $\{u,v\}$ lie on the base plane and they are determined by finding the four vertices of the quadrilateral containing grid G. The first two vertices are found as the pair of points G_a and G_b such that the distance between them is the largest among all points in G. The third point G_c is set equal to the point which has the largest distance from the line passing through G_a and G_b. The fourth point G_d is set equal to the farthest point from G_c.

The unit vector v is selected along the longer side of quadrilateral and u is chosen to be orthogonal to v and to preserve the right handedness of the original coordinate system (coordinate system for point cloud S). Finally, grid G is aligned with $\{u,v\}$.

73

4) Split G into subsets G_k

Rebar is often bent and also deflects under its own weight and other dead loads. Therefore, when modeling the rebar as straight cylinders, the rebar is divided up into shorter lengths as these sections can be more accurately represented as straight cylinders. The length of the sections is set to approximately 1 m so that there would be sufficient points to fit a cylinder center line and the bars were generally straight over this length. Therefore, grid G is divided into subsets G_k, for $k = 1$ to 6. For each G_k, the intersections of the rebar projected onto the base plane are determined in the following three-step procedure:

4a) Build histograms along u and v

Components $u_n = g_n \bullet u$ and $v_n = g_n \bullet v$ are calculated (where g_n is the n-th point from subset G_k) and their histograms are built, as shown in Figure 6-10. The bin size is set to $2R$.

4b) Segment rebar along u and v

The peaks in each of the histograms are identified and used to segment each individual rebar in G_k. Indices to points in the original data set, S, are used to obtain the segmented rebar in S.

4c) Fit cylinders and find intersections

A cylinder centerline is fitted to every segmented rebar section in S. The cylinder fitting was done by either minimizing the orthogonal or directional error function (see Section 7.2). The intersection of two centerlines in 3D is calculated as a point which has the smallest sum of the squared distances to both center lines. Then, all intersection points are projected along the drilling direction w onto the base plane.

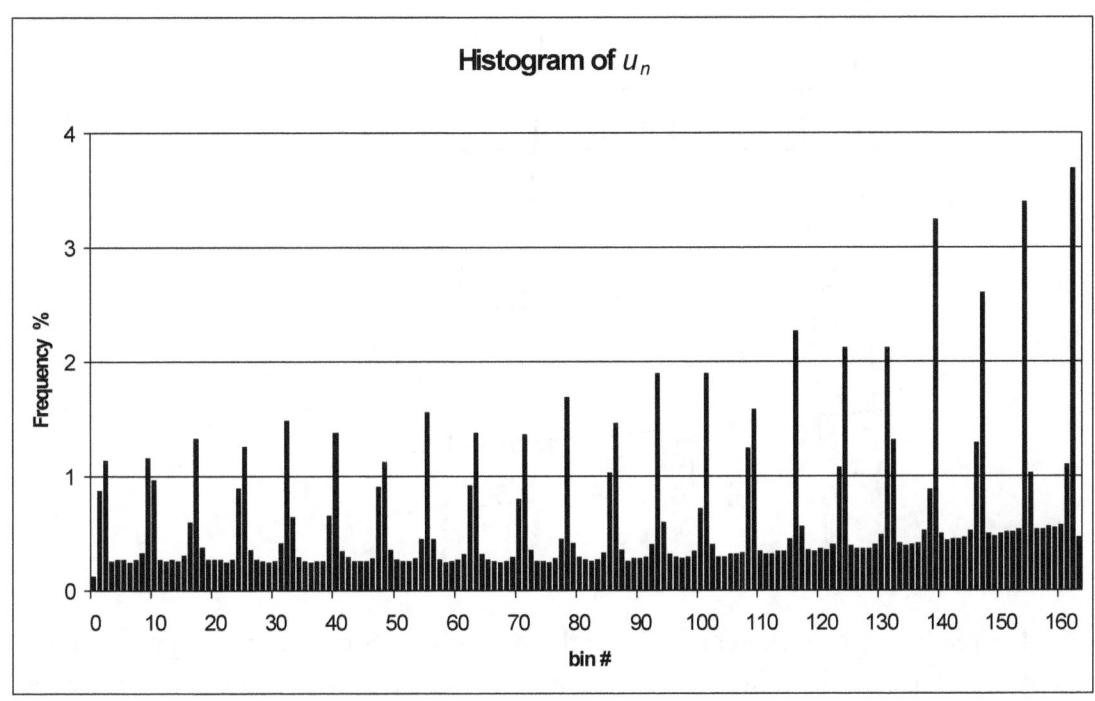

Figure 6-10 Histogram of the components of vectors g_n along direction u on the base plane. Individual peaks correspond to a single rebar. A similar histogram is built for components of g_n along direction v. The bin size is set to 2R.

5) Form quadrilateral cells

The rebar intersections obtained from the previous step, 4c, for all G_k are merged together and ordered to form a 2D grid according to their corresponding components along u and v. Quadrilateral cells are then created on the base plane, as shown in Figure 6-11, with an offset $Q = R$ from the rebar centerline.

75

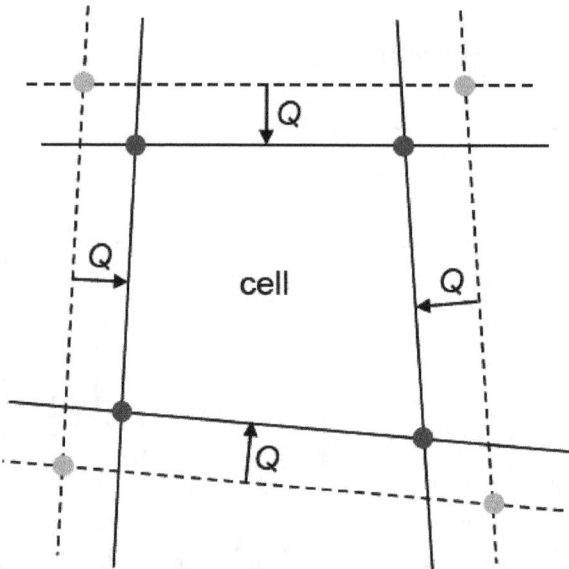

Figure 6-11 Creation of quadrilateral cells on the base plane. Straight lines represent the centerlines of the rebar (obtained in step 4c) and are indicated by dashed lines. The grey dots indicate rebar intersections. Solid lines mark location of dashed lines offset by Q towards the center of the cell. Black dots represent the vertices of a single quadrilateral cell on the base plane.

6) Build 2D histogram, get threshold T

All points P_n in the original point cloud S are transformed to the new coordinate system $\{u,v,w\}$ defined in step 3. The transformed P_n are projected along the drilling direction w onto the base plane and a 2D histogram is built. The size of the histogram bin is set to $2R$ x $2R$ (see Figure 6-12).

If a point is within the cell, then the corresponding bin in that cell is flagged as "non-empty" and the total number of points within the cell is calculated. Points which are outside of the cells (e.g., points on the rebar) are not included in the calculations. The total number of "non-empty" 2D bins (N_{BINS}) is determined as well as the total number of projected points in all cells (N_{PTS}). Then, a tolerance threshold T is defined as the ratio of N_{PTS} to N_{BINS}.

7) Get cell drilling depth

Based on the threshold T, a safe drilling depth for each "non-empty" 2D bin is calculated. The distance, d_n, from the drilling plane to a point, P_n, is calculated as follows:

$$d_n = D_{max} - (\mathbf{P}_n - \mathbf{B}_0) \bullet \mathbf{w}$$

6-3

where $\mathbf{B}_0$ is a point on the base plane defined in step 2.

For each "non-empty" 2D bin on the base plane, a histogram $H(d_n)$ is constructed where

$$0 < d_n < D_{max}$$

6-4

and the bin size for distances d_n is $2R$. Each bin in the histogram[6] $H(d_n)$ contains a count of all the points which are in the bin, see Figure 6-12. The location of the bin containing more than T points determines the depth of drilling for a given "non-empty" 2D bin, see Figure 6-12. If a 2D bin does not contain a bin with a count larger than T then D_{max} is assigned as the drilling depth for that bin; if a bin contains a count larger than T, then the drilling depth is set equal to the distance from this slice to the drilling plane. If two or more bins have counts larger than T, the drilling depth for the bin closest to the drilling plane is selected. Conservatively, the drilling depth for the entire cell is set equal to the smallest drilling depth from all the 2D bins within that cell. If the drilling depth of the cell equals D_{max} then the cell is labeled as safe. Otherwise, the cell is marked as unsafe.

---

[6] A 2D bin is inside a cell shown in Figure 6-11 and only points $\mathbf{P}_n$ which belong to the peak c) in Figure 6-9 contribute to the histogram $H(d_n)$.

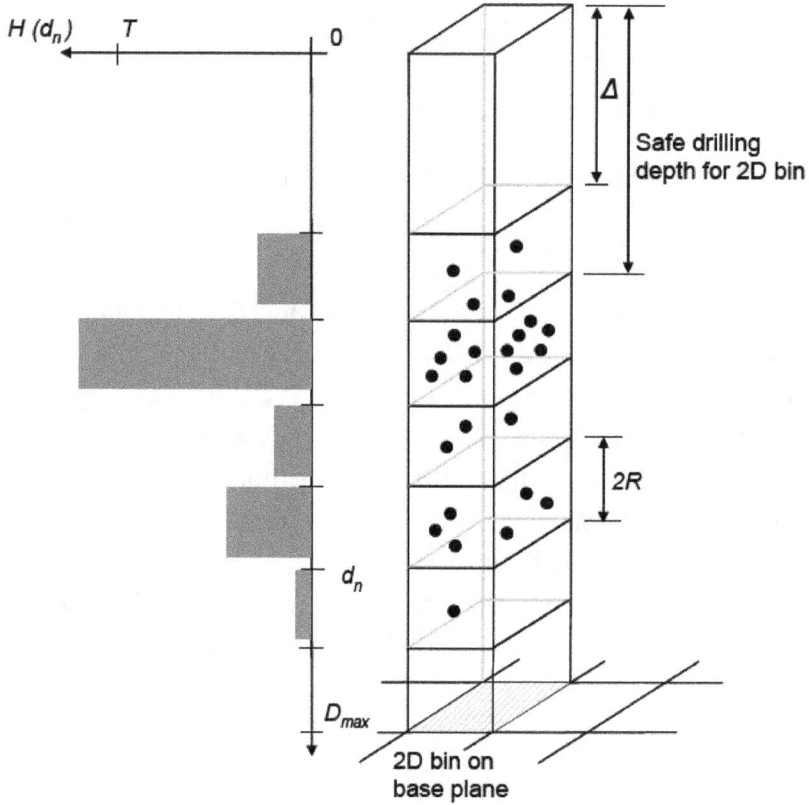

Figure 6-12 Determining the safe drilling depth for a single 2D bin on the base plane.

6.4 BIM Update Software

The results from the rebar intersection extraction algorithm are further processed in order to visualize the safe drilling zones in the BIM of the rebar cage. The four corners of each cell generated by the algorithm described in Section 6.3 are used to generate volumes in an IFC model and that model is used to update the BIM. The shape of the volume for safe drilling zones reflects the as-built rebar cage alignment. Volumes are only generated where it is permissible to drill through the entire depth of the rebar cage. Zones where it might be permissible to drill partially through the depth of the rebar cage are flagged as unsafe drilling zones and no volume is generated.

An example of a BIM of safe drilling zones is shown in Figure 6-13. The safe drilling zones are shown as grey volumes. The shape of the "top" of the volume is based on the four corners generated by the rebar intersection extraction algorithm. The safe drilling zones are shown in perspective in Figure 6-13, so for volumes located at the periphery, more of the "sides" of the volume are visible. The "height" of the volumes corresponds to the allowable drilling depth within the zone.

Note the variation of the size and shape of the safe drilling zones that reflects the as-built position and orientation of the reinforcing bars in the rebar cage. The absence of a safe drilling zone indicates the location of the three conduits in the rebar cage.

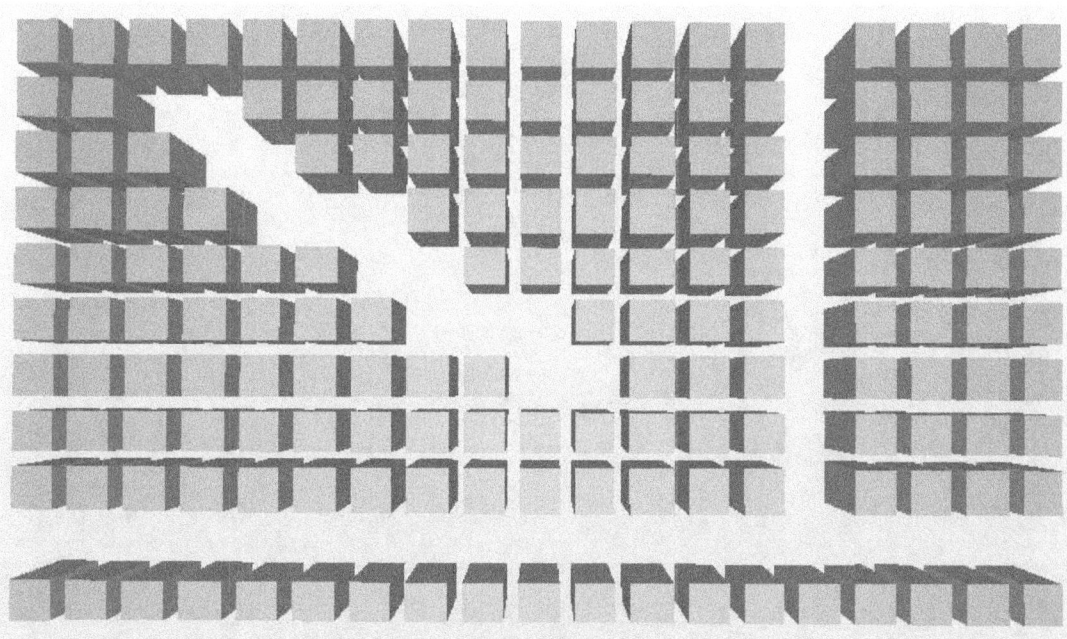

Figure 6-13 Safe drilling zones.

6.5   Laser Projector

The rebar intersection extraction algorithm also produced the required files that were used to display the safe and unsafe cells using the laser projector described in Section 5.3.8.2. This provided the audience with live feedback as to the accuracy of the technology. However, this process is analogous for field applications in which the locations of safe cells could be entered into a total station, which can be used to point to the safe drilling cells on top of the cured concrete.

7 Testbed Results

7.1 Approximate Times for Data Collection and Processing

7.1.1 Laser Scanning

The approximate times to obtain data of the rebar cage with the laser scanner for one scan location were:

1. Set-up instrument: 15 min
2. Conduct a preview scan (low density scan): 2 min
3. Conduct a high density scan of the region around the rebar cage: 10 min
4. Scan registration targets (8 total): 30 min

The total time for data collection was therefore about one hour per instrument setup or two hours altogether to capture the entire rebar cage. In general, this time will vary depending on the size of the area being scanned and the number of instrument setups required to cover that area. In addition, as shown in the above time results, the total time is also highly dependent on the number of registration targets used. In the case of the rebar cage scans, four to six targets would probably have been sufficient.

The times required to process the laser scanner data before it can be given to the rebar intersection finding algorithm were as follows:

1. Registration of the two scans: 2 min
2. Segmentation of the combined scan: 5 min
3. Exporting the data: 1 min

The total time for processing the laser scan data was approximately eight minutes.

7.1.2 D4AR Image-based 3D Reconstruction

The time required for the D4AR image-based 3D reconstruction technology has two components: (1) data collection time and (2) data analysis time. In the IACJS Testbed demonstration, it took approximately 45 min to capture 850 images of the rebar cage. The plan was to capture as many photos and as much details as possible to enable the bottom layer of the rebar along with the pipes to be fully observed and to enable the construction of a highly dense 3D point cloud. Nonetheless, the 3D point cloud had to be generated within a 48 hour time frame due to time constraints. Hence, among these images, only 380 images were randomly selected and processed. Prior to processing, the spatial resolution of these images was synthetically reduced to 2 megapixels and 1.2 megapixels for the SfM and MVS/VCL algorithms, respectively. The minimum number of required images for visual consistency control in both MVS and voxel coloring/labeling algorithms was set to three. Such visual consistency constraints reduce the chances of reconstructing those areas that are only visible from a maximum of two images, yet can result in a highly accurate 3D reconstruction.

Table 7-1 presents the breakdown of the computational time and algorithmic components required for generating a 3D point cloud with a density of 9 264 880 points. In this case, if the two megapixel images were used for the MVS/VCL algorithms, a 3D point cloud with five to ten times the density of the currently reconstructed point cloud could have been generated.

The total computational time for the SfM, MVS, and VCL algorithms was 32 hours and 29 min. This computational time was benchmarked on an Intel Core i7 3.2 GHz CPU with 12GB RAM on a LINUX platform. All modules in the SfM algorithm were sequentially computed, though the MVS and voxel coloring/labeling algorithms were computed using a parallel computing platform. A parallel implementation of the SfM can significantly reduce the computational time of the first step (expected computational time of about five minutes to ten minutes for the size of this dataset) (as experienced by Agarwal et al. (2009)). In addition, a multi-core GPU-CUDA based implementation can further minimize the overall data processing time. If a computer with higher memory capacity (RAM) and parallel computation were available, the total computational time could have been reduced to between 30 min and 60 min (not verified yet).

Table 7-1 Data processing results of the D4AR 3D reconstruction.

Number of images	380
Spatial resolution	2 Megapixel (approximately)
Number of images successfully registered	380
Point Cloud Density	9 264 880 points
Computational Time	32 hours 29 min

7.2   Results of the Rebar Intersection Extraction Algorithm

The rebar cage built for the demonstration consisted of 22 longitudinal and 13 transverse rebar which intersect at 286 locations. As mentioned in Section 5.3.2.1, the ground truth point cloud was obtained from only one scanner location which covered the area containing only 21 longitudinal and 13 transverse rebar. For this dataset, the total number of rebar intersections was 273. The rebar intersections as obtained in Section 5.3.2.1 are used as the ground truth rebar intersections.

The algorithm (Section 6.3) was able to identify the correct number of rebar intersections from all of the point clouds acquired using laser scanning and photogrammetry, where the input parameter R was set to 0.01 m (the radius of #6 rebar) and D_{max} to 0.279 m (which includes an offset $\Delta = 0.089$ m). The algorithm also determined the correct number of corresponding cells (240 for the ground truth point cloud data and 252 for all of the other point clouds, see Figure 7-1).

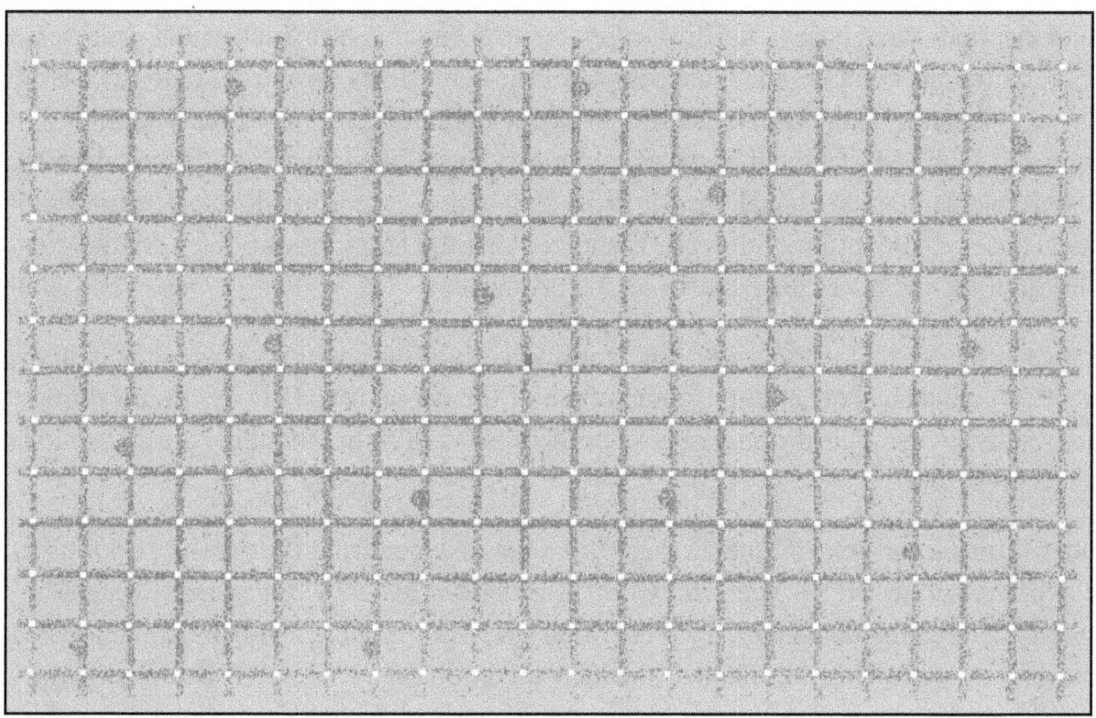

Figure 7-1 Top layer of rebars with overlaid rebar intersection points (white squares) calculated for Point Cloud 4 (Ptcld 4).

Four point clouds captured using the laser scanner and the photogrammetry/D4AR technology were processed through the rebar extraction algorithm and the rebar intersection were calculated and compared with the rebar intersections of the ground truth (obtain as described in Section 5.3.2.1). The four point clouds used were:

1. Ptcld 1: The point cloud from the ground truth instrument. The same point cloud was processed three times: a) manually, as described in Section 5.3.2.1, to obtain ground truth rebar intersections; b) using the algorithm described in Section 6.3 with an orthogonal error function; c) using the algorithm described in Section 6.3 with a directional error function.
2. Ptcld 2: The point cloud obtained from a laser scanner as described in Section 5.3.3. The rebar intersections were determined using the algorithm described in Section 6.3 with an orthogonal error function.
3. Ptcld 3: The point cloud obtained from the photogrammetry/D4AR technology as described in Section 5.3.4. The rebar intersections were determined using the algorithm described in Section 6.3 with an orthogonal error function.

83

4. Ptcld 4: The same as Ptcld 3 but reduced by a factor of 50 in order to approximate the condition in which fewer pictures are taken during data collection. The rebar intersections were determined using the algorithm described in Section 6.3 with an orthogonal error function.

The rebar intersections were based on fitting cylinders to the points (see Section 6.3, step 4c) and defining the rebar intersections as the intersections of the centerlines (projected onto the base plane) of these cylinders. Therefore, the ability to fit the cylinders correctly is essential. Preliminary results showed that the deviations as obtained using the orthogonal fitting algorithm for Ptcld 1 were worse than the deviations for Ptcld 2, 3, and 4. These results were contrary to expectations. Closer examination of the results showed that the larger deviations were due to incorrect fitting of the cylinders for Ptcld 1 (see Figure 7-2).

Figure 7-2 clearly shows that using the orthogonal fitting algorithm on this data results in an incorrectly fitted cylinder. A quantitative comparison of the two cylinders in the figure shows that they are relatively parallel (the angle between them is $0.277°$), but that they are 13.4 mm apart at their closest point.

Most cylinder fitting algorithms minimize the orthogonal distance from the measured points to a theoretical cylinder surface. It has been shown by Franaszek et al. (2009) that when fitting a sphere to noisy 3D imaging data taken from only one side of the sphere and using an algorithm which minimizes the orthogonal distance of the measured points to the sphere, calculating the center of the fitted sphere may result in more than one solution. The authors believe that a similar phenomenon occurs when fitting cylinders and may cause the fitted cylinder axes (and hence the calculated rebar intersections) to also be in the wrong locations. Figure 7-2 shows a point cloud of a section of rebar (approximately 1 m in length) from the rebar cage captured using the ground truth instrument. The figure also shows two cylinders, where the lower cylinder is the one fitted to the point cloud using a directional fitting algorithm and the upper cylinder is the one fitted to the point cloud using an orthogonal fitting algorithm. The view in the figure is along the axes of the lower cylinder. The points on top of the upper cylinder were captured from rebar that are perpendicular to and on top of the 1 m section of rebar.

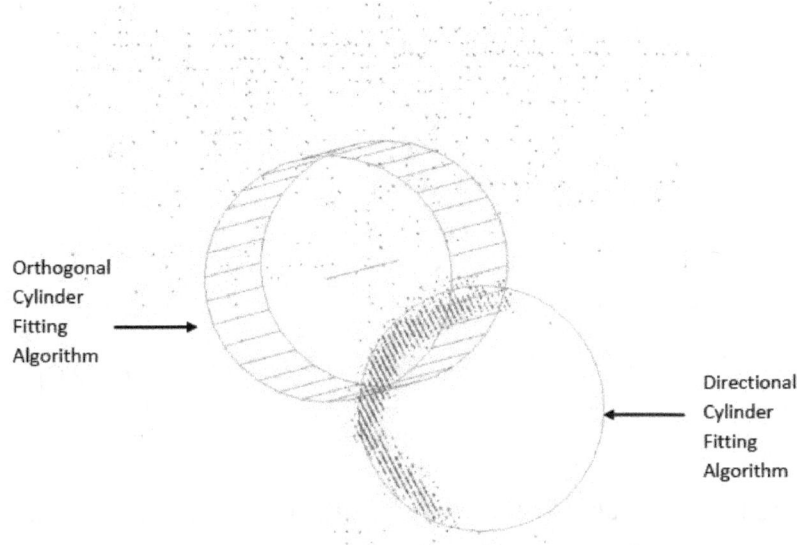

Figure 7-2 The difference between the orthogonal and directional cylinder fitting algorithms applied to a point cloud of a 1 m long rebar section captured with the ground truth 3D imaging instrument.

When the directional error function was used for Ptcld 1, the deviations were as expected (smaller than those for Ptclds 2, 3, and 4) and visual examination of the fitted cylinder with the point cloud showed that the fitted cylinder was in the expected location.

Ptcld 2 is the combination of two point clouds obtained from two locations and Ptclds 3 and Ptcld 4 also contain points from all sides of the rebar. In this case, the directional fitting, which works only for a point cloud acquired from one location with a line-of-sight instrument, was not used to fit the cylinders. Note that for Ptcld 2, the directional fitting could have been used for the two individual point clouds prior to registration.

Table 7-2 shows the average and standard deviation, maximum and minimum deviations of the rebar intersections from the ground truth rebar intersections for the four point clouds. Columns (1) and (2) show results for the same point cloud, Ptcld 1, processed with the orthogonal and directional error functions, respectively. The bottom row of Table 7-2 shows the total number of points in each dataset.

Table 7-2 Deviations from ground truth of the rebar intersections determined for the different datasets.

Deviation [mm]	(1) Ptcld 1 (orthogonal)	(2) Ptcld 1 (directional)	(3) Ptcld 2	(4) Ptcld 3	(5) Ptcld 4
Average / stdev	9.0 / 6.5	2.9 / 2.1	5.4 / 2.1	3.5 / 1.7	3.1 / 1.5
maximum	20.1	8.8	12.5	13.6	15.7
minimum	0.2	0.2	0.9	1.1	1.0
Number of points	762 747	762 747	469 695	6 942 603	138 853

The results in column (2) of Table 7-2 indicate that the rebar intersection extraction algorithm itself has an error of ± 2.9 mm on average. Therefore, the results in columns (3) to (5) could be improved if the rebar intersection extraction algorithm were improved, although the extent of the improvements in the numbers in columns (3) to (5) was not quantified and beyond the scope of this report.

7.3 Results of the Safe/Unsafe Prediction

As with any binary (i.e., true/false) classifier, the possible results of the safe/unsafe classifier are:

1. correct – identified a safe cell as safe and an unsafe cell as unsafe
2. false positive – identified an unsafe cell as safe
3. false negative – identified a safe cell as unsafe

Table 7-3 contains a summary of the cell status prediction results for the four point clouds discussed above. Shown in Figure 7-3 to Figure 7-6 are graphical depictions of the results in Table 7-3 as point clouds with the points from the lower rebar layer and the floor removed (a) and the corresponding cell status (b).

Table 7-3 Summary of cell status prediction.

	(1) Ptcld 1[*]	(2) Ptcld 2	(3) Ptcld 3	(4) Ptcld 4
# False positives	1	3	0	0
# False negatives	22	1	108	77
Total false predictions	23	4	108	77
Total # cells	240	252	252	252
[*] Ptcld 1 – using the directional error function for the cylinder fitting.				

These results do not characterize the performance of the technologies tested at finding safe and unsafe cells, but rather that of the rebar intersection extraction algorithm. A different algorithm may produce different results for each of the point clouds processed. These results indicate that the algorithm performs best with Ptcld 2 when false negatives need to be minimized. It is possible that Ptcld 1 produced more false negatives due to the fact that it is composed of only one scan from one side of the rebar cage, which may increase the ratio of points considered noise to points on the rebar.

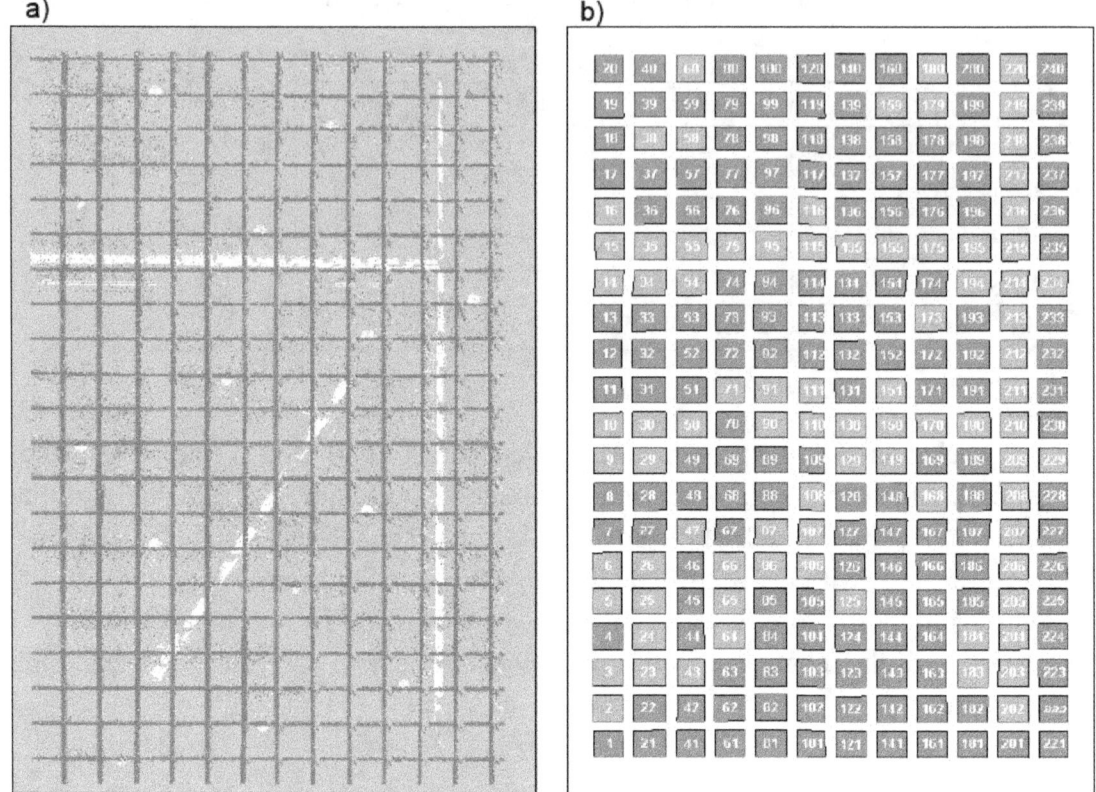

Figure 7-3 Prediction of cell status for Ptcld 1: a) top view of point cloud showing upper layer of rebar and objects (in white) - pipes between the upper and lower layer of rebars and targets mounted on the upper rebar layer; b) predicted cell status - red = safe, blue = unsafe. There were 22 false negative cells and one false positive cell.

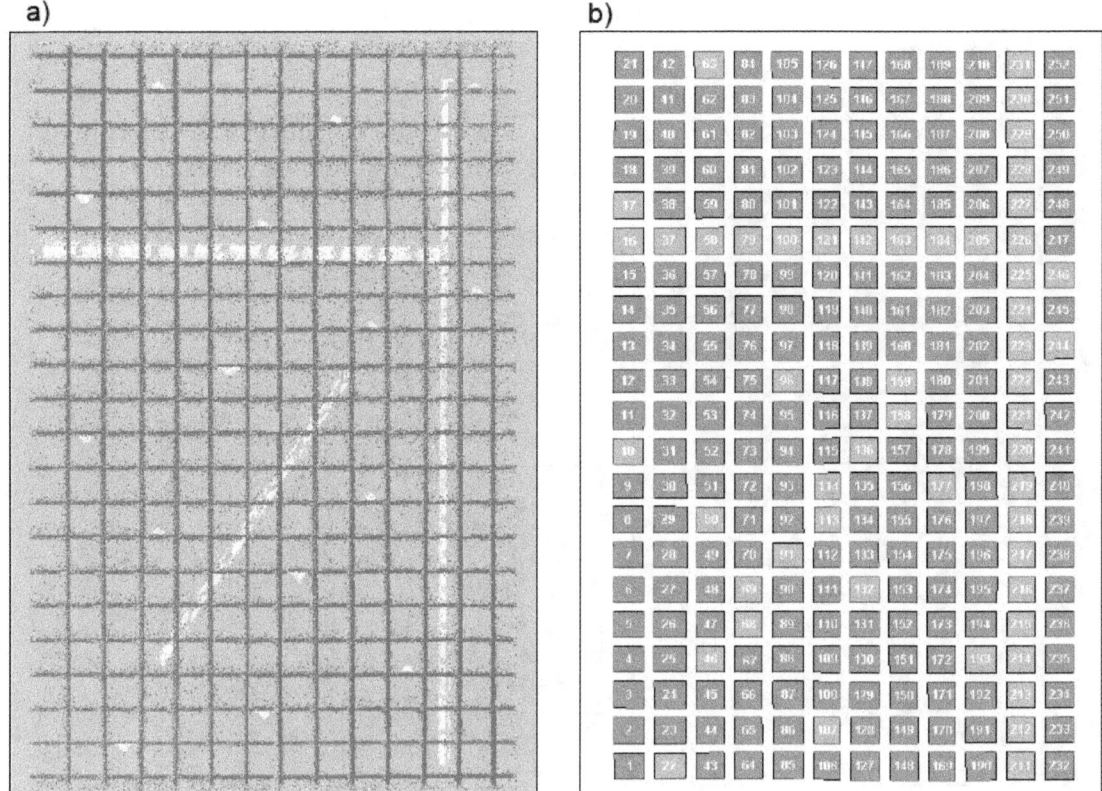

Figure 7-4 Prediction of cell status for Ptcld 2 (laser scanning): a) top view of the point cloud showing the upper layer of rebar and objects (in white) - pipes between the upper and lower layers of rebar and targets mounted on the top layer of rebar; b) predicted cell status – red = safe, blue = unsafe. There was one false negative cell and three false positive cells.

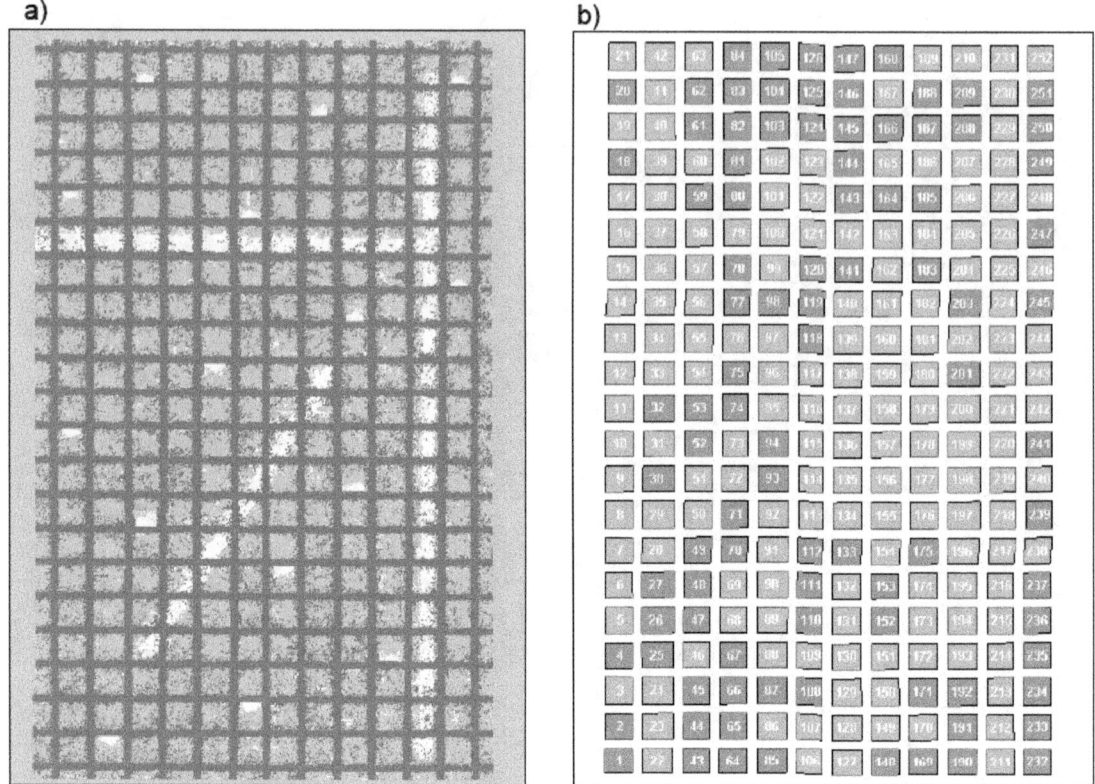

Figure 7-5 Prediction of cell status for dataset Ptcld 3 (photogrammetry/D4AR): a) top view of point cloud showing upper layer of rebar and objects (in white) - pipes between the upper and lower rebar layers and targets mounted on the upper rebar layer; b) predicted cell status – red = safe, blue = unsafe. There were 108 false negative cells and no false positive cells.

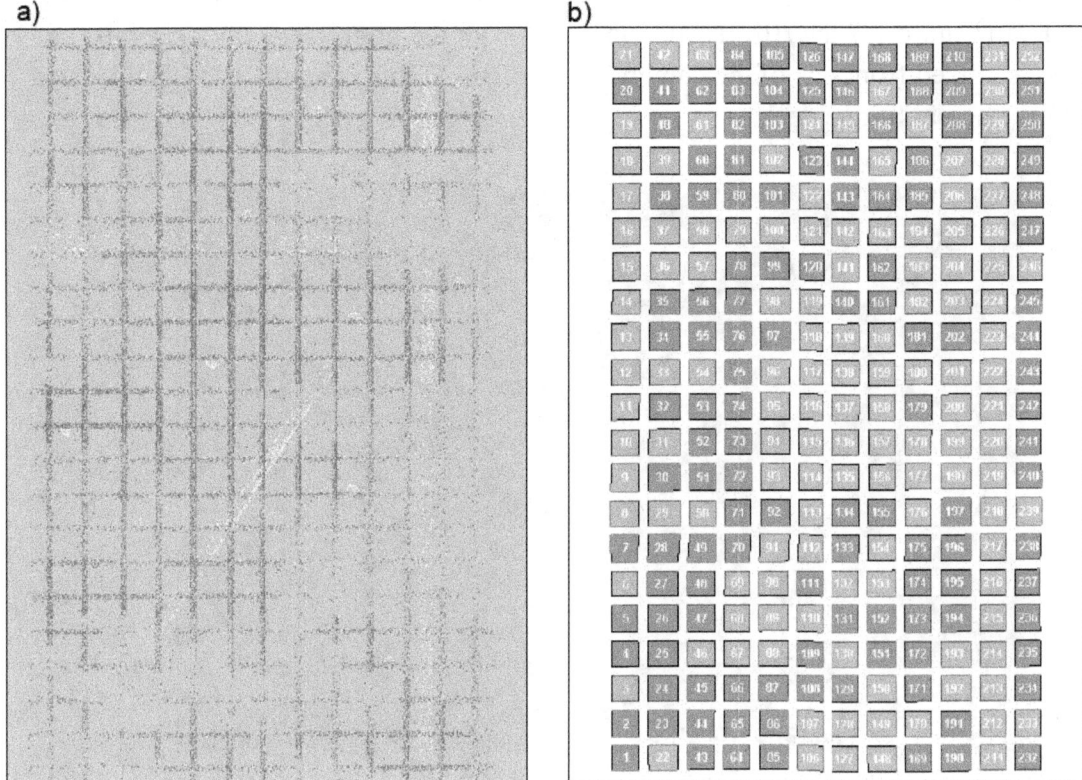

Figure 7-6 Prediction of cell status for dataset Ptcld 4 (photogrammetry/D4AR – decimated point cloud): a) top view of point cloud showing upper layer of rebar and objects (in white) – pipes between the upper and lower rebar layers and targets mounted on the upper layer of rebar; b) predicted cell status shown at a depth halfway between the drilling and the base plane. There were 77 false negative cells and no false positive cells.

The total time required to process Ptcld 1, Ptcld 2, and Ptcld 4 in order to calculate the rebar intersections and predict the safe/unsafe cells using the algorithm described herein was approximately 20 s per dataset. The time to process Ptcld 3 was approximately 90 s due to the large number of points in that dataset.

7.4 Visualization

7.4.1 D4AR Image-based 3D Reconstruction Results

Once the transformation of the point cloud to the iGPS coordinate system is computed, the BIM can be superimposed onto the transformed point cloud. For this purpose, the D4AR visualization component of Golparvar-Fard et al. (2009) is utilized. In this visualization component, several combinations of the point cloud, geo-registered imagery, and the BIM can be rendered together. As an example, in Figure 7-7 the BIM is superimposed with the point cloud and is semi-transparently visualized from one of the camera viewpoints. In this case, the BIM includes an IFC-model of the actual bridge deck.

Figure 7-7 The D4AR model visualized through a semi-transparent image.

Figure 7-8a and b visualize the integrated point cloud, BIM, and the geo-registered cameras. In Figure 7-8a, the geo-registered IFC-based BIM is visualized, while in Figure 7-8b the location and orientation of one of the cameras is highlighted. A user can navigate in the resulting augmented reality model by choosing various modes of visualization and conducting walkthroughs among semi-transparent imagery.

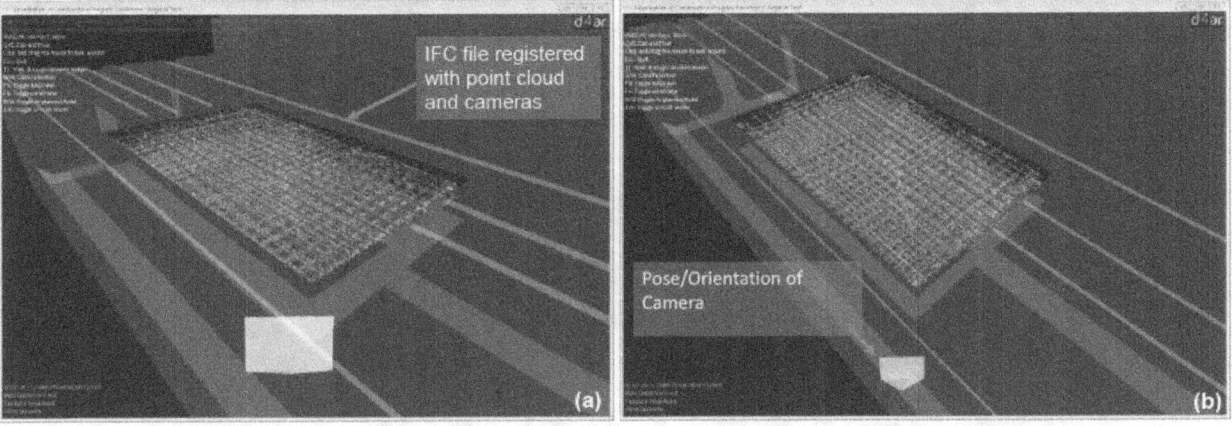

Figure 7-8 Visualization of the integrated point cloud, BIM, and geo-registered camera model. The location/orientation of one of the geo-registered cameras is highlighted in (b).

7.4.2   BIM Results

Safe and unsafe drilling zones computed from the rebar intersection extraction algorithm were visualized for the ground truth, laser scan, and photogrammetry/D4AR data. The safe drilling zones are displayed along with the ideal as-designed rebar cage model. In general, the rebar cage will appear between the safe drilling zones; however, due to variations of the rebar placement in the as-built rebar cage, sometimes there are clashes between certain safe drilling zones and the as-built rebar cage.

Figure 7-9, Figure 7-10, and Figure 7-11 show the safe drilling zones from the ground truth data. Drilling zones are not computed along two sides of the rebar cage (left, right, and top sides in Figure 7-10) due to limitations of the ground truth data. Notice the variation of the size and shape of the zones in Figure 7-11.

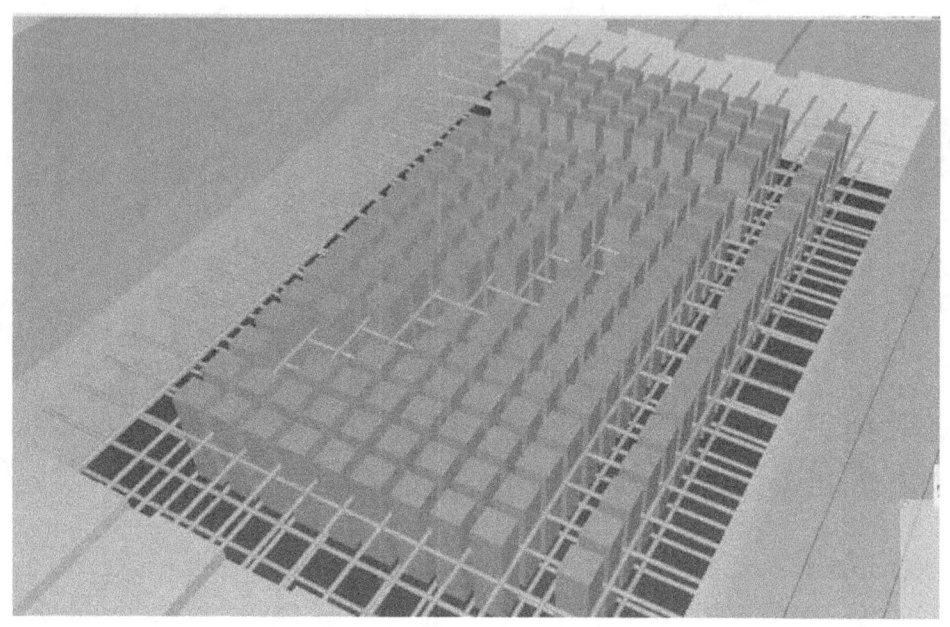

Figure 7-9 Perspective view of safe drilling zones computed from ground truth data.

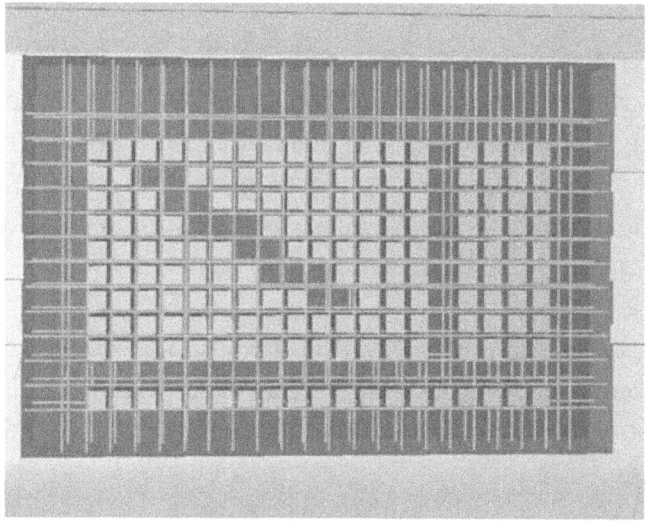

Figure 7-10 Overhead view of safe drilling zones computed from ground truth data.

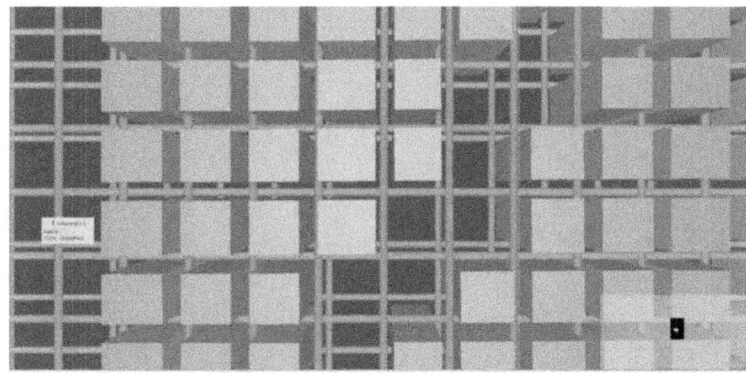

Figure 7-11 Close-up view of safe drilling zones computed from ground truth data.

Figure 7-12 shows the safe drilling zones computed from the laser scanner data. Comparing the pattern of safe drilling zones from the laser scan data with the ground truth data shows some significant differences. There are differences in safe drilling zones around the position of the diagonal conduit. The other differences in the safe drilling zones are due to rebar targets placed in the physical rebar cage which produce unsafe drilling zones.

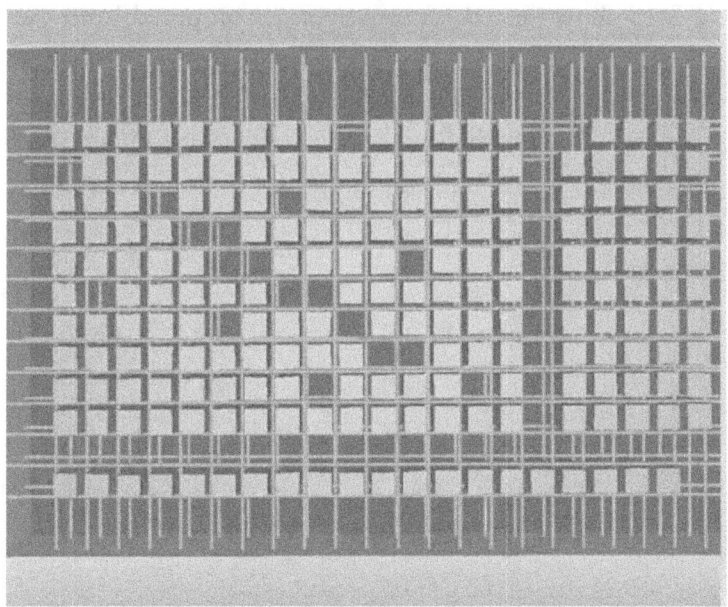

Figure 7-12 Overhead view of safe drilling zones computed from the laser scanner data.

Figure 7-13 shows the safe drilling zones computed from the photogrammetry/D4AR data. Some of the unsafe zones are due to missing data in this dataset. Note the significant variation in the pattern of safe drilling zones from the laser scan and ground truth data.

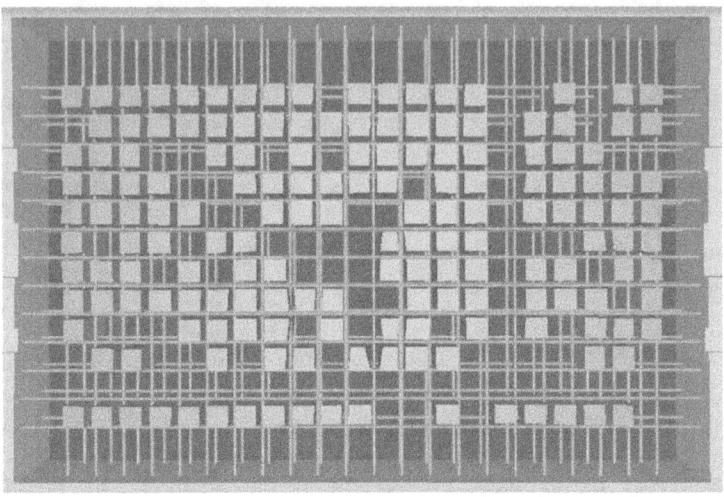

Figure 7-13 Overhead view of safe drilling zones computed from photogrammetry data.

Figure 7-14 and Figure 7-15 show the safe drilling zones computed from the ground truth data as voids in the bridge deck model, i.e., the reverse of showing the safe drilling zones as volumes in the previous figures. Everything that is blue is an unsafe drilling zone. The green reinforcing bars that are visible in the as-built safe drilling zones indicate the differences between the ideal as-designed rebar cage and the position and orientation of the reinforcing bars in the as-built physical model of the rebar cage.

Figure 7-14 Overhead view of safe drilling zones as voids computed from ground truth data.

Figure 7-15 Perspective view of safe drilling zones as voids computed from ground truth data.

Figure 7-16, Figure 7-17, Figure 7-18, and Figure 7-19 show a comparison of the safe drilling zone volumes between the ground truth and laser scan data. The volumes for both data are shown together in the visualizations where the ground truth safe drilling zones are shown in red and the laser scanning safe drilling zones are shown in yellow. Zones that are all yellow show safe drilling zones that are computed from the laser scan data that are not in the ground truth data, particularly along the diagonal conduit.

Figure 7-18 shows the slightly different safe drilling zone geometry computed from each dataset. The clash between the green reinforcing bars with the safe drilling zones in Figure 7-19 shows the differences between the as-designed and as-built reinforcing bar alignment.

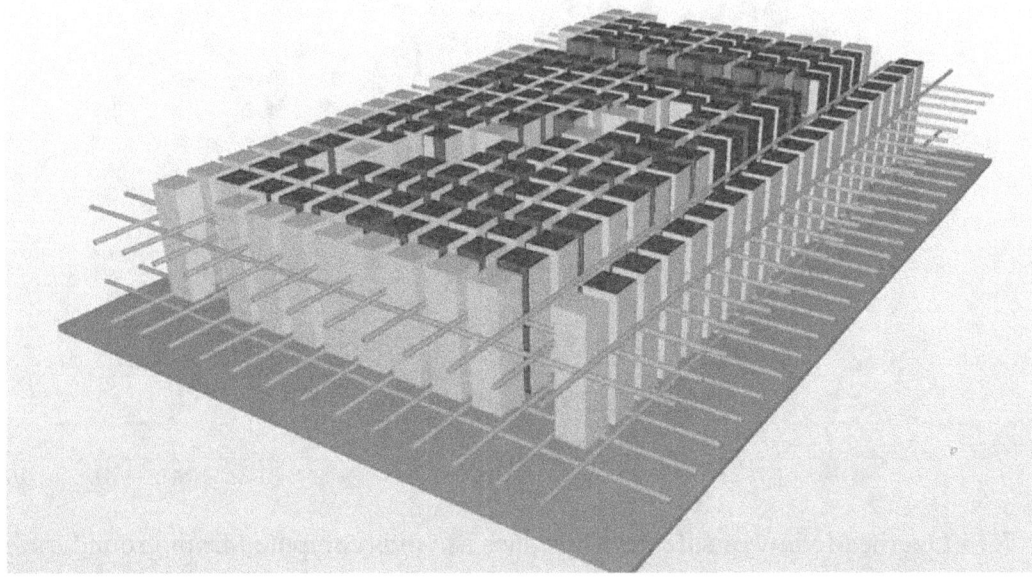

Figure 7-16 Perspective view of safe drilling zones comparing ground truth and laser scan data.

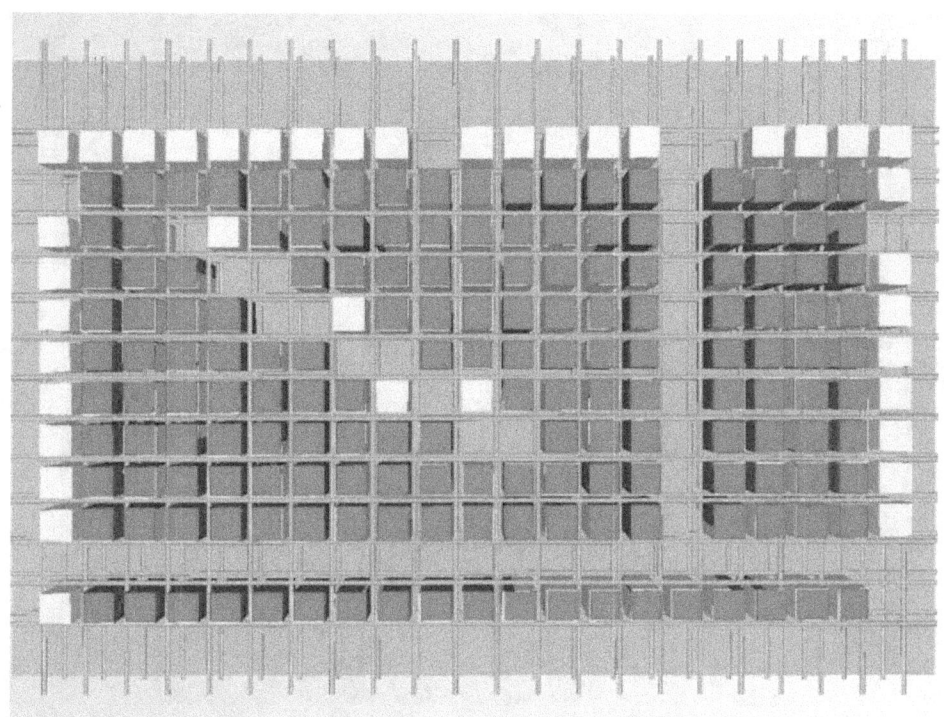

Figure 7-17 Overhead view of safe drilling zones comparing ground truth and laser scan data.

Figure 7-18 Close-up view of safe drilling zones comparing ground truth and laser scan data.

Figure 7-19 Close-up view of safe drilling zones comparing ground truth and laser scan data showing overlap with reinforcing bars.

7.5 Limitations of the Results

7.5.1 Data Homogeneity

The point density for the various datasets varied greatly (see Table 7-2) due to the different technologies used. In addition, the data captured with the ground truth instrument was captured from only one side of the rebar cage while the data gathered with the laser scanner was captured from two sides. Therefore, in general the laser scanner data may be considered more uniform than the ground truth data since it includes information from two sides of the rebar. Furthermore, the point density is not spatially uniform within each dataset that is captured using a laser scanner since objects closer to the scanner will generally include more points than those further away. Therefore, neither the laser scanner data nor the ground truth data of the rebar is homogeneous across the rebar cage.

The D4AR pictures were collected manually. Therefore, the percent overlap between the pictures and the amount of rebar detail captured is not uniform either.

7.5.2 Registration

In order to compare the laser scanner and photogrammetry/D4AR results with each other and with ground truth, all the data had to be registered to the IACJS Testbed reference frame (the iGPS coordinate system).

The D4AR data was registered by manually selecting, in the point cloud, the centers of the rebar targets. Since the centers of the rebar targets had already been measured with the iGPS, the D4AR data could be transformed into the iGPS coordinate system. However, because the manual selection of the targets is a subjective process, the error caused by this registration step may be unpredictable.

In contrast, the laser scanner scans (two) were automatically registered together using registration targets mounted on the walls of the testbed. Since the registration targets had already been measured with the iGPS, the laser scanner data could also be transformed into the iGPS testbed reference frame. Although this registration step was less subjective because of the use of the registration targets, the entire process when repeated three times produced different registration errors. This may be because in each of these three times, the least square minimization algorithm starts from a slightly different initial guess of the solution and because the registration targets were manually weighted differently each time in order to minimize the registration error.

7.5.3 Uncertainty Analysis

The uncertainty in the 3D position measurements from the iGPS and the ground truth instrument are ± 250 µm and ± 100 µm, respectively. The uncertainties in the data obtained from the laser scanner and the camera hardware and the D4AR software combination are unknown. The uncertainties in the various plane and cylinder fits conducted on the data captured during the experiments were also not available. The uncertainties for the fitted parameters (e.g., the uncertainty of the location of the cylinder centerline) can be estimated by repeating the data collection and data processing several times or by sub-sampling one dataset several times. However, this was beyond the scope of the demonstration. Nevertheless, for a given technology, the uncertainty of the deviations from ground truth of the rebar intersections can be estimated as the standard deviation of the mean from all the calculated intersections (see Table 7-2).

Furthermore, it is also unknown how the registration process affects the uncertainty of the data. Methods of quantifying these errors (i.e., how registration errors propagate into measured data) have not been investigated.

7.5.4 Rebar Intersection Extraction Algorithm

The rebar intersection extraction algorithm presented in Section 6.3 assumes that all of the rebar in the input datasets (point clouds) have the same diameter and that the diameter is known a priori. Although this may be true for certain field conditions (it was true for the rebar cage in the testbed experiments), in general a rebar cage may consist of multiple rebar sizes with different diameters.

The above algorithm also assumes that once the concrete is poured on top of the rebar the resulting concrete surface will be a flat plane parallel to the floor of the rebar cage and that any drilling will be perpendicular to that finished concrete surface. The algorithm would have to be extended in order to account for other than normal drilling angles and for a more complex finished concrete surface.

7.5.5 Simulated Environment

Finally, the results from the demonstration described in this report are also limited by the fact that all the experiments were conducted in a clean and controlled laboratory environment using a mockup of a rebar cage. Actual field conditions are often more congested and cluttered and the site is much larger than the laboratory setting. Congestion and size of the site lead to longer set-up and data collection times. Congestion also leads to data occlusions due to objects (e.g., people, equipment, tools, general clutter) being in the way and limited accessibility to locations for setting up the instruments or taking photos.

Rebar cages also often contain objects or items that introduce noise to the data. Items such as rebar ties, rebar chairs, formwork, trash, and tools are often found in or on rebar cages under actual field conditions. Rebar hooks, bends, and splices will also add complexity. The ability to handle the added noise and complexity may require the development of a better point cloud filtering algorithm and a more sophisticated classifier.

8 Productivity Analysis: Original Method versus the 3D Imaging Method for the Rebar Mapping Scenario

8.1 Analysis Strategy

The strategy for the productivity analysis is to compare the time for the original method where the dowels are prepared, attached to the planks, and removed after concrete placement (original method) versus mapping the rebar locations using either laser scanning or photogrammetry (3D imaging method). The analyses presented herein are meant to identify the maximum duration of the 3D imaging methods in order to provide a productivity advantage over building the rail bridge spans using the original method. The analyses do not differentiate whether laser scanning or photogrammetry is used in obtaining the 3D data, as the method does not impact the productivity.

The following assumptions are made:

- A deck paving machine could not be used in either method because the wood planks were necessary to make the recesses on the surface.
- Estimates in this report are based on one typical railway span, as seen in Figure 8-1.

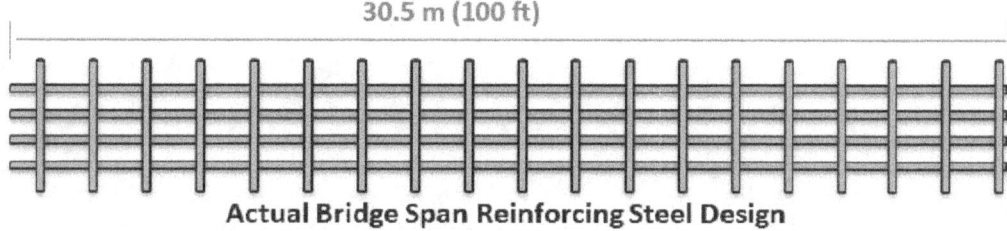

Actual Bridge Span Reinforcing Steel Design

Figure 8-1 Typical reinforcing steel mat used on a railway bridge span

8.2 Methodology

To calculate the potential time savings achieved from using either 3D imaging method, a discrete event simulation (DES) modeling approach was used. In DES modeling, a process can be modeled with logical relationships with time as a factor. Each task to be performed can be associated with prior task completion or the release of a dependent resource for that task.

Several simulation systems have been developed for typical construction processes. EZStrobe was selected for this analysis since it allows for creation of a network based on activity cycle diagrams and time constraints (through clock advance and event generation mechanisms based on activity scanning or three-phase activity scanning) (Martinez 2001). Figure 8-2 outlines the DES model developed in EZStrobe for the original field method of installing the wooden dowel rods for the embedments. Appendix A contains a list of variable definitions, as the software truncates and restricts certain characters.

Railway Pavement Construction Process with Wood Blockouts

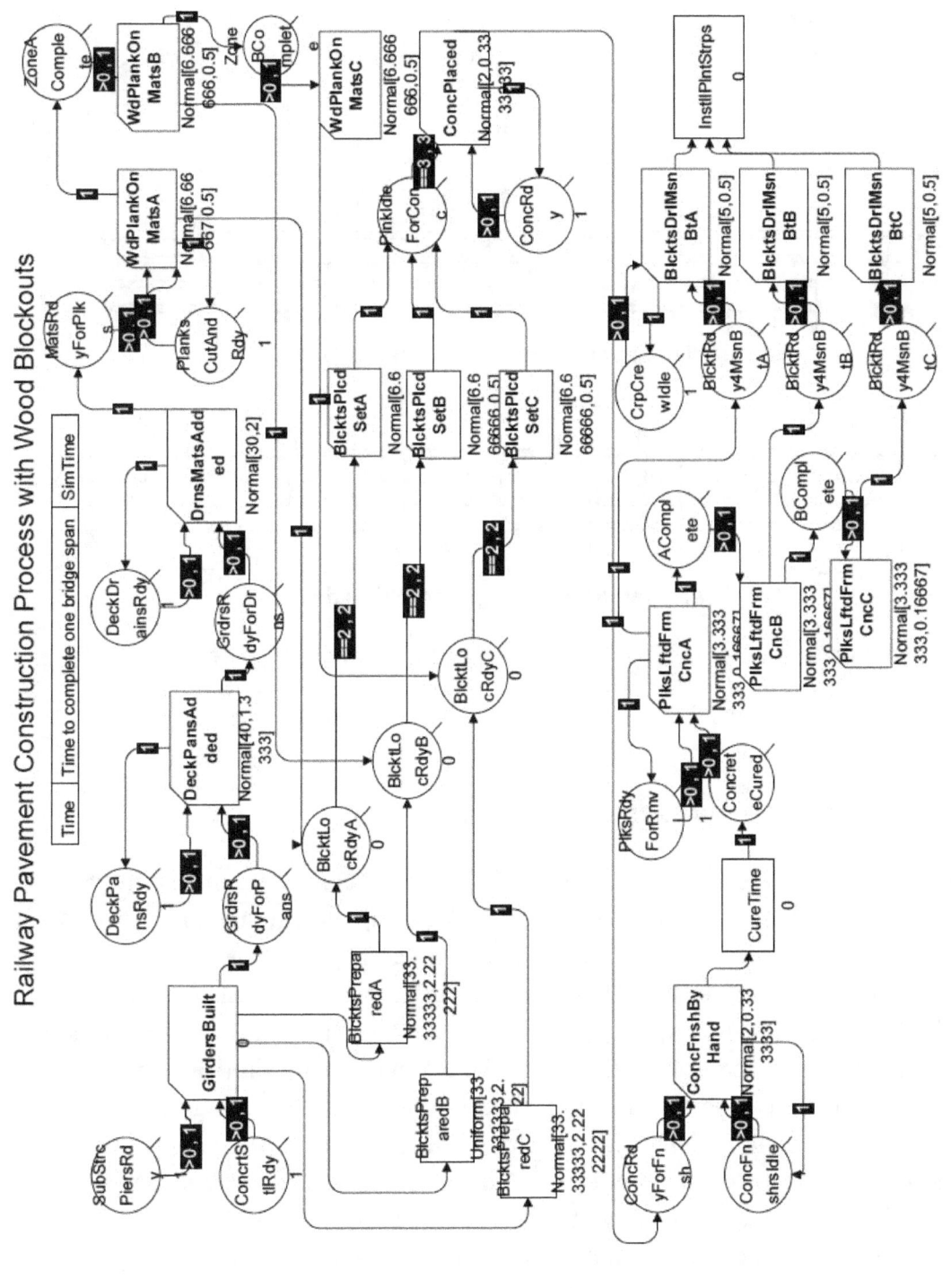

Figure 8-2 EZStrobe DES model for the original method.

105

The model logic was derived from several visits to the jobsite and discussions with managers, engineers, and foremen on the project. For this study, each span was divided into three zones; A, B, and C (see Figure 8-3). This allows for parallel processes that were coordinated by the management staff. The times, in hours it takes for individual activities, were also obtained during the site visits. The durations in the DES model (Figure 8-2) used a normal frequency distribution with ranges that incorporate six standard deviations from the mean (three each way).

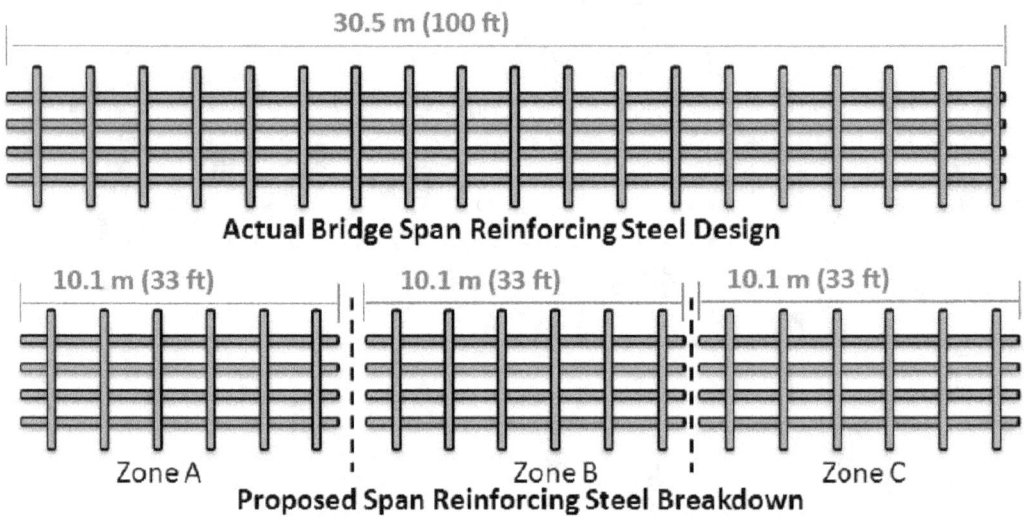

Figure 8-3 Breakdown of bridge span for parallel processes.

8.3 Breakeven Analysis and Results

The DES models provide a breakeven analysis for the use of a 3D imaging method in the construction of a typical railway bridge span. After proper setup, the original model was run over 100 iterations to develop an average duration for all activities. That average duration then became the target total duration for the 3D imaging model. Logical durations for activities involved in the 3D imaging method were input at a level that resulted in an overall duration similar to that of the original method. At this point, the two models are similar in total duration, and the breakeven analysis can occur. The time it takes for activities involving the 3D imaging method were summed to identify the breakeven point. The times in the original method, as previously discussed, are true to the judgments of managers and field engineers on the project. For the DES model for 3D imaging method, the tasks and durations for 3D imaging practices were used in place of installing the blockouts used in the original method. Essentially, the tasks in the original method that would not be performed in the 3D imaging method are eliminated. They are replaced by 3D imaging tasks such as equipment setup, image acquisition, and processing of the data, which could be required irrespective of whether laser scanning or photogrammetry was used.

The original method (using wood dowels as blockouts) results in an average duration of approximately 115 hours (or 11.5 days based on the project's standard 10 hour work days) and a standard deviation of 2.90 hours. The model was set with a parameter of 100 iterations to achieve a significant sample size. The fastest possible time the model suggests is 109 hours, while the slowest time is 123 hours. This information is summarized in Table 8-1. A detailed breakdown of the results can be seen in Appendix B. To help clarify results, the DES original model is reproduced as a process diagram in Figure 8-4 with the durations for each task shown as well.

Table 8-1 Summary of the original method's simulation results.

Method (100 iterations)	Average (hours)	Standard Deviation (hours)	Minimum (hours)	Maximum (hours)
Original (blockout)	115.46	2.897	109.15	123.34

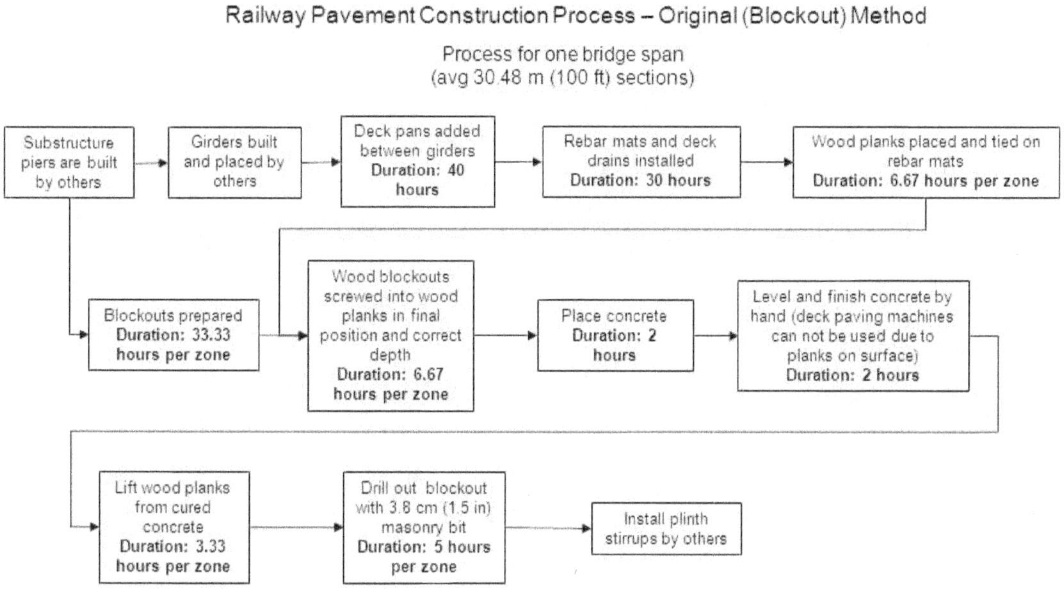

Figure 8-4 Process flow diagram for the original (blockout) method.

With an average of 115 hours using the original method for a typical 30.48 m (100 ft) bridge span, the researchers modified the time for the activities related to the 3D imaging method to reach that time threshold. The resulting DES model can be seen in Figure 8-5. Appendix C

107

provides a list of the variable names and their definitions. The 3D imaging method eliminates activities that are associated with wood blockouts in the original method. Preparing the blockouts, placing and attaching them to the wood planks, and removing/drilling out the blockouts are no longer necessary in the 3D imaging model. The added steps involved are measuring targets in the site coordinate system (SCS), setting up the equipment, measuring registration targets, acquiring data, moving and repeating for a full model, processing the data, and finding and drilling out the locations for the dowel sleeves.

Railway Pavement Construction Process with 3D Imaging

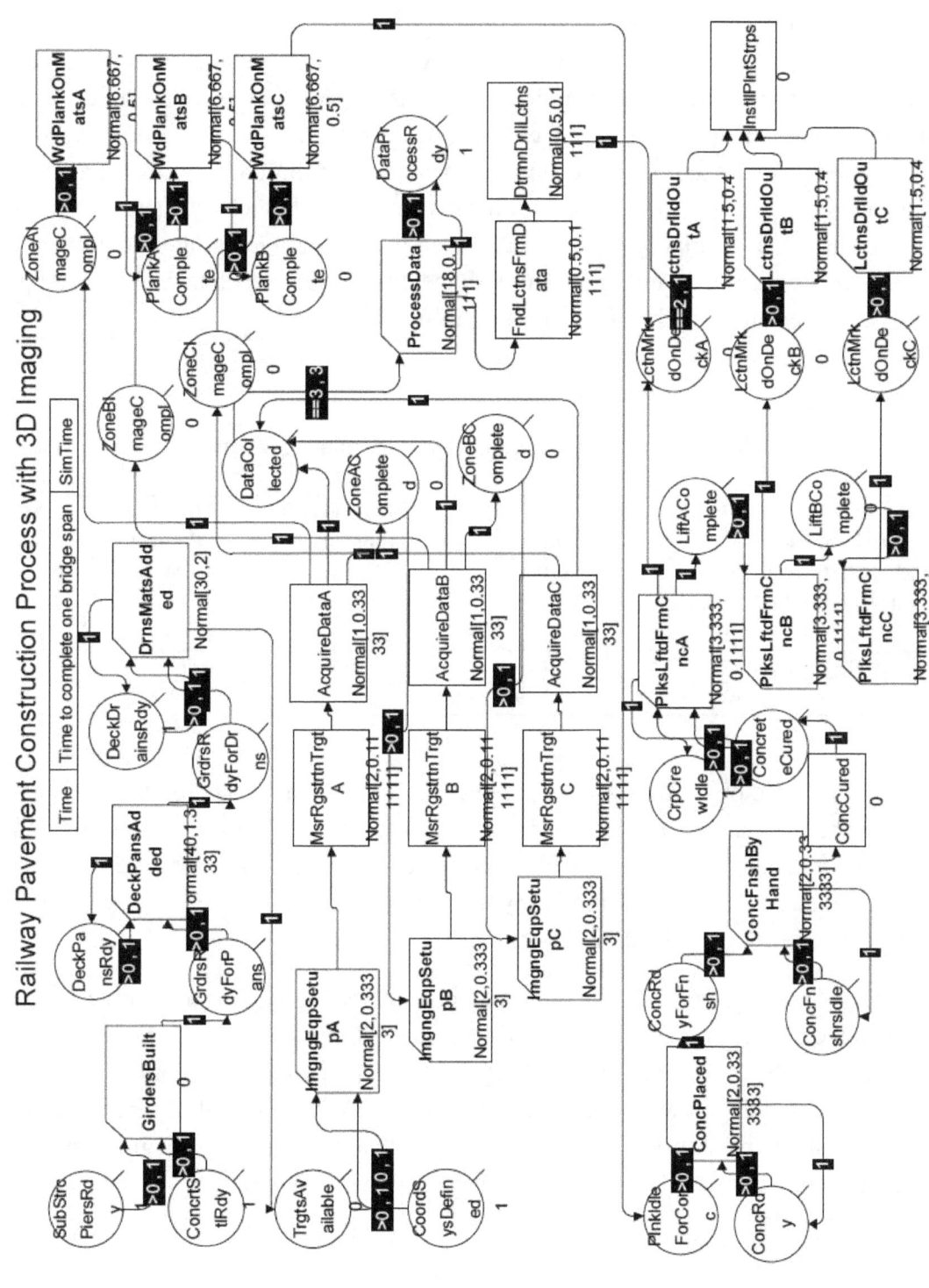

Figure 8-5 EZStrobe DES model for the alternative method (i.e., using 3D Imaging).

109

The 3D imaging model produces an average of 110 hours and a standard deviation of 2.61 hours. Similar to the original method DES model, the software ran 100 iterations of the process and produced a minimum of 104 hours and maximum of 116 hours (see Table 8-2). Table 8-2 also reports the total time for image acquisition and image processing from the model (based on one model iteration) as 14.19 hours and 18.09 hours, respectively. The software limits the ability to view report statistics of activities when multiple iterations are run, therefore, the image acquisition and processing durations were based on one observation. Since the figures were based on only one observation, there is no standard deviation, minimum, or maximum to report. The numbers are similar to the original method results using wood blockouts, since all are within 4 % of the original value. Similar to the original model, the DES 3D imaging model is reproduced as a process diagram in Figure 8-6 with the durations for each task shown. Appendix D contains detailed results from the 3D imaging DES model.

Table 8-2 Summary of 3D imaging simulation results.

Alternative 3D Imaging Method (100 iterations)	Average (hours)	Standard Deviation (hours)	Minimum (hours)	Maximum (hours)
3D Imaging Method (Overall Model including construction and 3D imaging activities)	110.09	2.614	104.56	116.00
Image Acquisition (1 observation)	14.19	N/A	N/A	N/A
Image Processing (1 observation)	18.09	N/A	N/A	N/A

From the durations based on field data from the described project and logic built into the 3D imaging model, the 3D imaging activities for a single 30.48 m (100 ft) bridge span (including image acquisition and data processing) would need to occur within a range of 21.5 hours to 38.5 hours. This is due to data processing not lying on the critical path because of parallel processes. The processing can take up to 18 hours.

By comparing the durations in the 3D imaging model to a model based on the rebar cage in the IACJS Testbed, the durations for individual activities are within a reasonable expected range. In the Testbed scenario, the time to measure and register targets, set up the equipment, and acquire the data all occurred within a few hours. In the DES model, the time for all of those activities to occur was 15 h total, or five hours per zone. The activity that may cause issues in a breakeven analysis is the data processing. This is discussed further in Section 8.4.

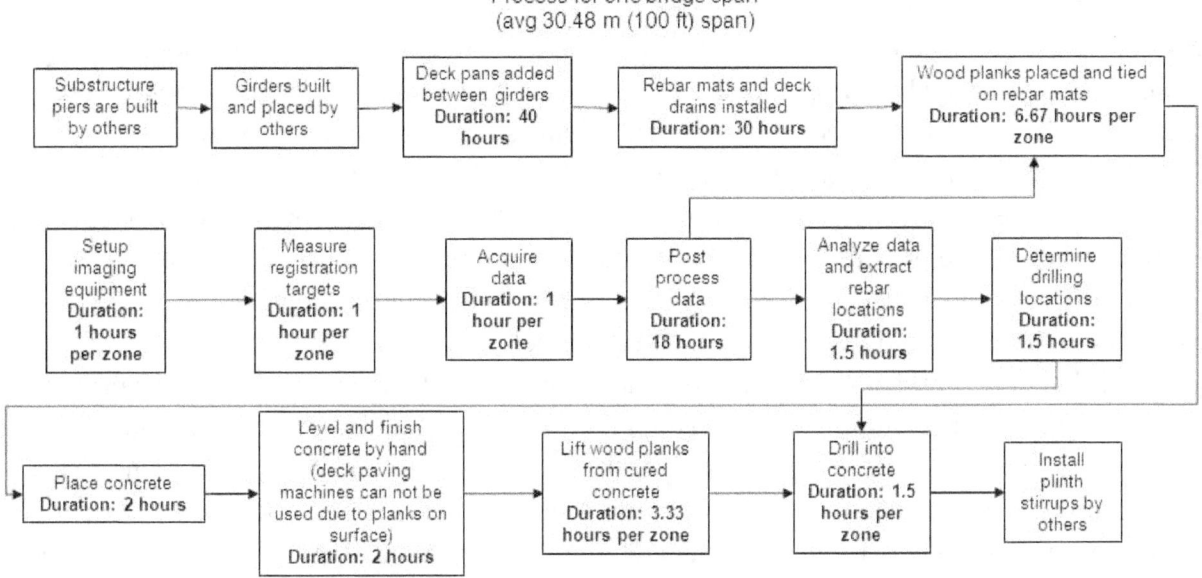

Figure 8-6 Process flow diagram for 3D imaging method.

8.4 Discussion

While the 3D imaging method provides a more reliable approach than the wooden dowel approach, there are some drawbacks suggested by management and craft at the jobsite. The main issue is that setting up and acquiring images for a 3D model while the reinforcing steel is exposed could potentially delay the craftsmen from placing the concrete. A proposed process for acquiring the 3D image would be for field crews to first install all of the steel reinforcement mats for a single bridge span and then allow the engineers or surveyors to take the images. After the image acquisition and processing is complete, the field crews would commence with placing the concrete. The craftsmen and foremen found the proposed process to be disruptive and would likely decrease worker motivation.

Possible solutions to this issue include parallel work, off-shift image acquisition, and acquiring the 3D imaging data from airborne equipment. Subsequently, the models previously presented implemented parallel processes by splitting up each span into three zones. This allows for minimum wait time for the craft while the 3D imaging data is being collected. If the crews still have to wait, the 3D imaging data acquisition can be easily scheduled prior to or at the end of the normal work day.

Another potential issue is the amount of time that would be required for processing the data. Figure 8-3 provided a simplified visualization of the 3D imaging process for the steel

111

reinforcement mats. The results of applying the photogrammetry/D4AR technology to the rebar cage in the IACJS Testbed required approximately 32 hours to process 380 images (see Section 7.1.2). The rebar cage consisted of a 2.44 m x 3.66 m (8 ft x 12 ft) rebar mat, which is smaller in layout than the mats required for the actual rail project. Through linear interpolation, the time required to process the bridge span images at the same scale as the IACJS Testbed would be approximately 320 hours per zone. That figure is unreasonably high, however, the IACJS Testbed images produce an extremely high quality output with redundancy in image acquisition that may be unnecessary for the actual project. It is likely that significantly fewer images of the reinforcing mat could provide the necessary data. A more reasonable and desirable level of quality needs to be established. To help provide a benchmark for this objective, Figure 8-7 shows the estimated durations based on the DES models that would need to be obtained for each process using either laser scanning or photogrammetry technologies on a 10 m (33 ft) bridge span section in order to minimize crew disruption. The image acquisition per zone can take up to five hours in duration, while the processing of the data can last up to six hours per zone. This would allow processing of Zone A to begin before field work starts on Zone B. The same logic applies to Zones B and C. It is assumed that there would be minimal effort to stitch the images of Zones A, B, and C together to create a cohesive image of the steel mat for the overall bridge span. In summary, image acquisition and processing for a typical bridge span that can be done within 33 hours (three sections total at 11 hours each for image acquiring and processing) provides a time-saving alternative to the original (blockout) method. In terms of the proposed D4AR solution, the next logical step would be to identify how many images could be obtained and successfully processed within the timeframe for each bridge section and determine if the resulting 3D image is accurate enough to identify locations for the embeds.

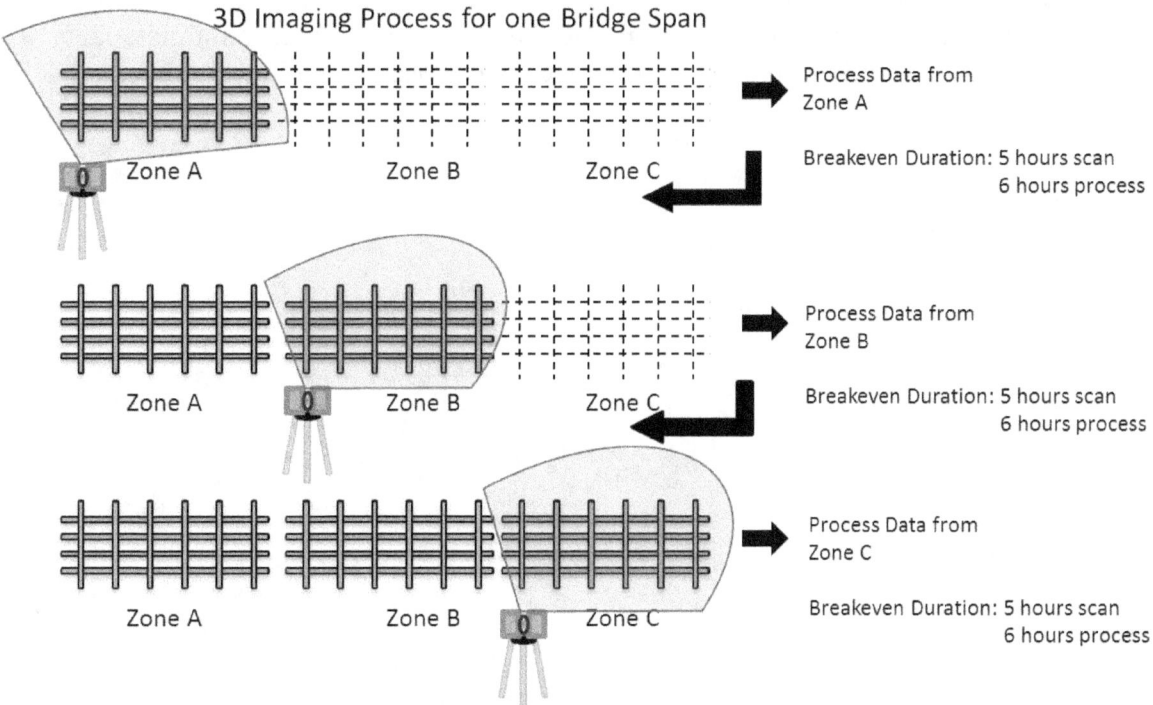

Figure 8-7 Simplified visualization of the 3D imaging process.

8.5 Summary

The methodology included in this section provides a process for analyzing a proposed alternative construction method using discrete event simulation (DES) models. By setting the total duration for the original method using wood blockouts and a 3D imaging method using either laser scanning or photogrammetry, a breakeven analysis was performed. Activities in the 3D imaging method must be completed within the durations identified in the model.

The proposed 3D imaging method provides a reasonable substitute for the original blockout method for determining embedment locations within a rebar mat. However, the challenge of successfully acquiring and processing a 3D image without delaying the placement of the concrete by a construction crew exists. The management team has critical decisions to make if the 3D imaging method is adopted; specifically the sequencing of the imaging process and the desired quality of the rebar mat model. The results of the analyses include the following:

1. If an entire steel reinforcement mat for a typical 30.48 m (100 ft) bridge span on the proposed railway bridge deck is photographed and processed at once, the photographing and processing would need to occur within 33 hours in order for 3D imaging to provide any benefit over the existing wood block out method. While this may reduce the entire duration for the construction of a single bridge span, it would still create undesirable wait times for the construction crews.

113

2. If a 30.48 m (100 ft) section is photographed in three sections, thus allowing processing and field construction to occur in parallel and thereby minimizing wait times, each 10 m (33 ft) section would need to be photographed and processed within 11 hours.

9 Summary and Recommendations for Future Research

9.1 Summary

The Intelligent and Automated Construction Job Site (IACJS) Testbed is a large (15 m x 16 m x 10 m) laboratory facility at the National Institute of Standards and Technology (NIST), in Gaithersburg, MD. The objectives of the testbed in fiscal years 2006 to 2011 were to enable innovation in and evaluation of new construction methods, equipment, and technologies; to provide a demonstration environment for these new technologies; to evaluate technology readiness and mitigate risk; and to enable the deployment of new technologies to the field. The creation of the testbed was motivated by the need for such testbeds that was identified through numerous industry initiatives.

The capabilities of the testbed include various equipment (such as mobile and cable-driven robots), sensors (such as laser scanners and laser trackers), and computation and visualization resources (such as high-end workstations and high-resolution projectors and cameras).

The application of 3D imaging technologies to six construction industry scenarios, which were identified in collaboration with industry partners, was selected for potential demonstration in the testbed. One of the scenarios, rebar mapping for cast-in-place concrete railway bridge decks, was selected as the initial construction activity to which 3D imaging would be applied. In this scenario the location of the rebar after the concrete is poured is needed in order to place embeds into the decks, which will support the train rails. Drilling into the concrete decks must not result in structural damage (i.e., the rebar must be avoided).

One of the current methods for placing the embeds involves placing blockouts (wooden dowels) into the rebar cage at the required embed locations prior to concrete pouring and subsequently removing them after the concrete has set. This method is labor intensive, prone to problems, and time consuming. Two alternative methods that make use of 3D imaging were evaluated in the IACJS Testbed. Both methods involve obtaining a 3D point cloud of the rebar cage before the concrete is poured and then relating that data to the site coordinate system. The data can then be used to mark the drilling locations that are rebar-free. One of the methods used a laser scanner to acquire the 3D data while the other method used a digital camera and custom software to process the 2D camera images into a 3D point cloud.

A reconfigurable rebar cage (R2C) was designed and fabricated and a custom algorithm was developed to process the resulting 3D data from the two technologies. The algorithm calculated the rebar intersections of the top layer in the rebar cage as well as the safe drilling zones and depths within the cage. Ground truth data about the locations of the rebar intersections within the rebar cage was obtained using a coherent laser radar (a type of laser scanner) with an uncertainty of $\pm$ 100 μm. Ground truth data about safe and unsafe drilling zones was obtained visually. The performance of the above two 3D imaging technologies was compared based on the deviations of the calculated rebar intersections from ground truth and based on the number of safe vs. unsafe zones identified by the above custom algorithm.

Furthermore, an analysis of the productivity of the original method (using wooden dowels) for placing the embeds into the concrete decks was conducted using Discrete Event Simulation techniques and the ASTM E2204 Standard Guide for Summarizing the Economic Impacts of Building Related Projects (ASTM, 2011). The alternative 3D imaging method was also analyzed using the same techniques and maximum times (in order to improve productivity over the original method) required to acquire and process the 3D imaging data were calculated based on the above analysis.

Finally, the two 3D imaging technologies and the methods that were developed to evaluate them were demonstrated in the IACJS Testbed to an industry audience on June 30, 2011.

9.2   Recommendations for Future Work

9.2.1   Data Acquisition

The application of the two technologies described in this report to rebar mapping for concrete railway bridge decks was a novel approach that the authors had not seen before. Therefore, the demonstration that took place in 2011 was only an initial implementation of these techniques to the rebar-mapping problem. Optimized methods for acquiring and processing the data needed in order to identify the rebar-free zones are required in order to improve the measurement and economic performance of these techniques.

In particular, the laser scanner used to acquire data from the rebar cage may not be the optimum type of laser scanner for this application. Newer, handheld scanners that do not require registration targets and that can acquire and process the data faster may be more suitable. The authors have conducted preliminary tests on such an instrument and the results are encouraging. Figure 9-1 shows various views of data that were acquired using a new handheld scanner on the rebar cage in the IACJS Testbed. The use of these types of scanners for this application needs to be further investigated.

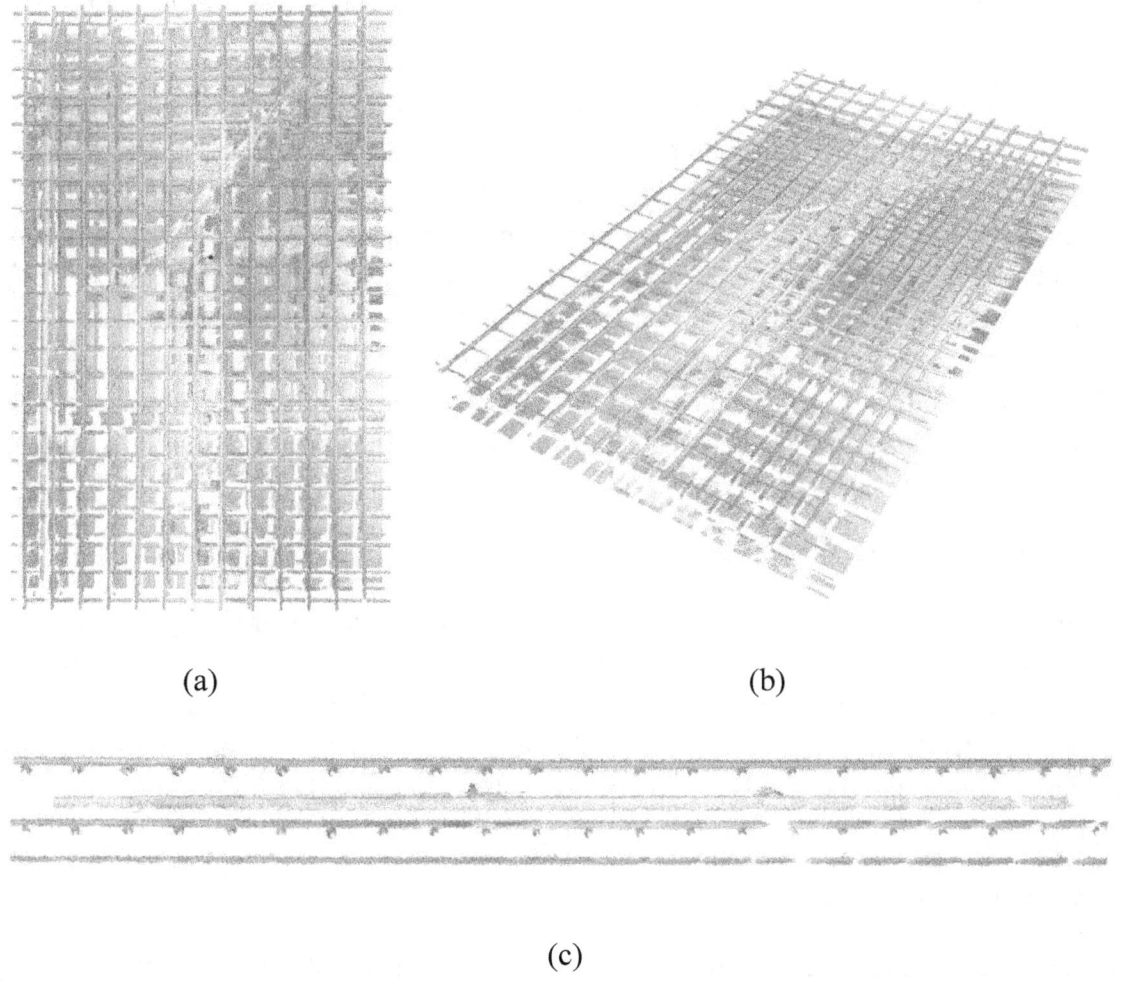

(a)   (b)

(c)

Figure 9-1 Top (a), perspective (b) and side (c) views of data of the rebar cage acquired using a new handheld scanner.

In addition, the 2D images of the rebar cage that were acquired and processed using the D4AR technology described in Section 5.2.2.2 are not optimized in terms of their resolution, the type of lens used, the angles from which the images were taken, and the number of pictures that were processed. The authors set all of the above parameters based on conventional practice from applications of photogrammetry in other fields, but further research is needed in order to optimize them for the rebar-mapping application.

117

9.2.2 Rebar Intersection Extraction, Safe vs. Unsafe Zone Identification and Drilling Depth Calculation Algorithm

The algorithm that was developed for the purpose of extracting the rebar intersection from the 3D point cloud data and for identifying the safe and unsafe drilling zones and depths was not optimized for this application. It makes certain assumptions that may not accurately reflect conditions in the field. However, this algorithm was not intended as a technological solution to the rebar-mapping problem, but rather as a research tool to be used for demonstration purposes. Further research would be required in order to improve its performance for actual field deployment. Specifically, the algorithm would need to be able to handle multiple rebar diameters, more complex rebar arrangements (such as non-parallel rebar layers and more than two rebar layers) and more noisy data (such as due to the presence of rebar ties, tools, and debris on and within the rebar cage).

9.2.3 Performance Metrics and Methods

The performance metrics and methods that were developed and demonstrated in the IACJS Testbed need to be implemented using field data from an actual construction site. The authors have already acquired field data from a railway construction project, but that data has not yet been processed due to time constraints. Processing this data and applying the same performance metrics that were used in the testbed will help refine the metrics further for actual field deployment.

9.2.4 Ground Truth

The ground truth instrument and methods used in the IACJS Testbed are not suitable for experiments with large rebar cages or for experiments conducted in the field due to the long times involved in acquiring the data and to the high cost and sensitivity (to environmental conditions) of the equipment. Techniques for acquiring ground truth data that are cheaper and more efficient would be needed for conducting experiments under more realistic (or actual) field conditions.

9.2.5 Standard Image and Point Cloud Datasets

As previously mentioned, the rebar intersection extraction algorithm would need to be improved in order to detect rebar-free zones in more complex rebar configurations that reflect common field conditions. Toward such an effort, standardized point clouds of common rebar cage configurations could be developed (with known ground truth models of the rebar) so that researchers could improve the algorithm or develop new ones. In the same vein, standardized digital images of the same rebar cages could also be acquired and made available for researchers who are developing photogrammetric solutions to rebar mapping problems.

9.2.6   Future Testbed Research

As mentioned at the outset of this report (Section 1.2), the objectives of the IACJS Testbed were to enable industry innovation, evaluation, demonstration, and deployment of new technologies for a broader range of construction challenges than just the rebar mapping scenario. NIST management has also recognized the applicability of the testbed's capabilities to industries beyond construction and in fiscal year 2012 is focusing resources on applications of the testbed to the manufacturing industry. In order to reflect that change in scope, the testbed was renamed to the Intelligent Sensing and Automation Testbed (ISAT). Near-term research within the ISAT will focus on developing performance metrics and characterization methods for manufacturing automation technologies (such as robots and machine vision sensors) and a means of assigning technology readiness levels for those technologies.

References

Agarwal, S., Snavely, N., Simon, I., Seitz, S., & Szeliski, R. (2009). Building Rome in a Day. *Proc., Int. Conference on Computer Vision.* Kyoto, Japan.

Akin, O., & Garrett, J. (2006). *FINAL REPORT - NSF-NIST workshop on Advanced Building Information Testbeds (NABIT).* Washington, DC: National Science Foundation.

Albus, J. S., Bostelman, R. V., & Dagalakis, N. (1992). The NIST ROBOCRANE. *Journal of Robotics System, 10*(5), 709-724.

ASTM (2011). E 2204-11 Standard Guide for Summarizing the Economic Impacts of Building Related Projects. West Conshohocken, PA: ASTM International. Retrieved from http://www.astm.org/Standards/E2204.htm

Boehler, W., Vicent, M. B., & Marbs, A. (2003). Investigating Laser Scanner Accuracy. *XIXth CIPA Symposium.* Antalya, Turkey.

Bosche, F. (2010). Automated recognition of 3D CAD model objects in laser scans and calculation of as-built dimensions for dimensional compliance control in construction. *Advanced Engineering Infomatics, 24*, 107-118.

Bouguet, J. Y. (2010). "Camera Calibration Toolbox for Matlab." Retrieved Nov 14, 2011 from http://www.vision.caltech.edu/bouguetj/calib_doc/.

Chapman, R. E., & Butry, D. T. (2008). *Measuring and Improving the Productivity of the US Construction Industry.* National Institute of Standards and Technology. Gaithersburg: NIST White Paper.

Debevec, P. E., Taylor, C. J., & Malik, J. (1996). Modeling and rendering architecture from photographs: A hybrid geometry and image-based approach. *Proceedings, SIGGRAPH conference*, (pp. 11-20).

Dobers, T., Jones, M., & Muttoni, Y. (2004). Using a Laser Scanner for the Control of Accelerator Infrastructure During the Machine Integration. *IWAA Workshop.* Geneva, Switzerland.

"Electromagnetic Induction Explained." GSSI. Retrieved Nov 4, 2011 from http://www.geophysical.com/whatisem.htm.

FIATECH. (2003). Retrieved from Capital Projects Technology Roadmap: http://www.fiatech.org

Franaszek, M., Cheok, G., Saidi, K., & Witzgall, C. (2009). Fitting Spheres to Range Data from 3-D Imaging Systems. *IEEE Transactions on Instrumentation and Measurement, 58*(10).

Furukawa, Y., & Ponce, J. (2006). *High-Fidelity image based modeling.* University of Illinois.

Furukawa, Y., & Ponce, J. (2010). Accurate, dense, and robust multi-view stereopsis. *IEEE Trans. Pattern Analysis and Machine Intelligence, 32*, 1362-1376.

Furukawa, Y., Curless, B., Seitz, S. M., & Szeliski, R. (2009). Reconstructing Building Interiors from Images. *Proc., Int. Conf. on Comput. Vision.*

Golparvar-Fard, M., Pena-Mora, F., & Savarese, S. (2009). D4AR - A 4-Dimensional Augmented Reality Model for Automating Construction Progress Data Collection, Processing and Communication. *J. of Information Technology in Construction, 14*, 129-153.

Golparvar-Fard, M., Pena-Mora, F., & Savarese, S. (2010). D4AR - 4 Dimensional augmented reality - tools for automated remote progress tracking and support of decision-enabling tasks in the AEC/FM industry. *Proc., The 6th Int. Conf. on Innovations in AEC Special Session - Transformative machine vision for AEC.* State College, PA.

Golparvar-Fard, M., Pena-Mora, F., & Savarese, S. (2011). Monitoring Changes of 3D Building Elements from Unordered Photo Collections. IEEE workshop on Computer Vision for Remote Sensing of the Environment (in conjunction with ICCV-11).

Goucher, S. (2010). Integrate High Definition Surveying Directly into Real-time Construction Site's Geomatics. Austin, TX: FIATECH. Retrieved from http://fiatech.org/webinar-archives/480-june-22-integrate-high-definition-surveying-directly-into-real-time-construction-sites-geomatics.html

Greaves, T. (2006). Modular Construction with Dimensional Control - Solutions for Tight Labor Market. *Sparview, 4.*

Greaves, T., & Hohner, L. (2009). *Benefits and Paybacks for Industrial Plant Design, Construction and Operation.* Spar Point Research.

"Ground Penetrating Radar Explained." GSSI. Retrieved Nov 4, 2011 from http://www.geophysical.com/whatisgpr.htm.

Hartley, R. I., & Zisserman, A. (2000). *Multiple View Geometry in Computer Vision.* Cambridge University Press.

Hartley, R., & Zisserman, A. (2004). *Multiple View Geometry.* Cambridge, UK: Cambridge University Press.

Hiremagalur, J. e. (2007). Creating Standards and Specifications for the Use of Laser Scanning in CALTRANS Projects. University of California at Davis.

Horn, B. (1987). Closed-form Solution of Absolute Orientation using Unit Quatemions. *J. of the Optical Society, A(4)*, 629-642.

Huang, A. L., Chapman, R. E., & Butry, D. T. (2009). Metrics and Tools for Measuring Construction Productivity: Technical and Empirical Considerations. *NIST Special Publication*.

IAEA-TCS-26. (2005). International Atomic Energy Agency, Training Course Series 26. Non-destructive testing for plant life assessment. Vienna, Austria.

Lichti, D. D. (2006). Error modeling, calibration and analysis of an AM-CW terrestrial laser scanner system. *ISPRS Journal of Photogrammetry and Remote Sensing, 61*, 307-324.

Longstreet, B. (2010). Scanning PDX. *Professional Surveyor Magazine*, pp. 20-23.

Lowe, D. (2004). Distinctive image features from scale-invariant keypoints. *Int. J. of Computer Vision, 60*(2), 91-110.

Lytle, A. M., Saidi, K. S., Stone, W. C., & Scott, N. A. (2003). Towards an Intelligent Job Site: Status of the NIST Automated Steel Construction Test Bed. *Proceedings of the 20th International Symposium on Automation and Robotics in Construction (ISARC)*. Eindhoven, Netherlands.

Martinez, J. (2001). EZStrobe - General Purpose Simulation System Based on Activity Cycle Diagrams. *Proceedings of the 2001 Winter Simulation Conference*, (pp. 1556-1564).

Moore, J. F. (1992). *Monitoring building structures.* NYC: Nostrand Reinhold.

National Research Council. (2009). *Advancing the Competitiveness and Efficiency of the US Construction Industry.* Washington, DC: National Academy Press.

Nister, D. (2004). An efficient solution to the five-point relative pose problem. *IEEE Trans. on Pattern Anaysis and Machine Intelligence (PAMI), 26*(6), 756-770.

PhotoModeler (2011). "Measuring and Modeling the Real World". Retrieved July 2011 from http://www.photomodeler.com/index.htm.

Presley, L., & Weber, B. (2009). Modularization Containment Vessels in New Nuclear Power Plants. *Power, 153*.

"Rebarscope[TM] Advanced System for Rebar Location and Bar Size Determination." James Instruments. Retrieved Nov 4, 2011 from http://www.ndtjames.com/Rebarscope-Complete-System-p/r-c-4.htm.

Riley, G. (2003). *Rebar Locators, How do they Measure Up?* Structure Magazine.

Salo, P., Jokinen, O., & Kukko, A. (2008). On the calibration of the distance measuring component of a terrestrial laser scanner. *The International Archives of the Photogrammetry, Remote Sensing and Spatial Information Sciences*, (pp. 1067-1071). Beijing.

Schoonmaker, S. J. (2010). Construction jobsite 4-D real-time modeling data and methods for standardized crane motion simulation and study. *Master's Thesis, Lehigh University*.

Slattery, D. K., & Slattery, K. T. (2010). Evaluation of 3-D Laser Scanning for Construction Applications. Illinois Center for Transportation.

Snavely, N., Seitz, S. M., & Szeliski, R. (2006). Photo tourism: exploring photo collections in 3d. *Proc. ACM SIGGRAPH '06.* New York.

Sternberg, H., & Kersten, T. P. (2007). Comparison of Terrestrial Laser Scanning Systems in Industrial As-Built Documentation Applications. In *Optical 3-D Measurement Techniques VIII* (Vol. I, pp. 389-397). Zurich: Gruen/Kahmen.

Takayama, J. (2009). 3D Image Reconstruction of Reinforcing Bar and Pipe Using Handy Micorwave Radar Scanning System. *ICROS-SICE International Joint Conference.* Fukuoka International Congress Center, Japan.

Triggs, B., McLauchlan, P., Hartley, R., & Fitzgibbon, A. (1999). Bundle adjustment - a modern synthesis. *Intl Workshop on Vision Algorithms.* Corfu, Greece.

Uffenkamp, V. (1993). State of the art of high precision industrial photogrammetry. *Proc., 3rd Int. Workshop on Accelerator Alighnment.*

Voegtle, T., Schwab, I., & Landes, T. (2008). Influences of different materials on the measurements of a terrestrial laser scanner (TLS). *The International Archives of the Photogrammetry, Remote Sensing and Spatial Information Sciences*, (pp. 1061-1066). Beijing.

Whitfield, C. R. (2010). 3D Scanning: Embrace Mobile 3D Laser Scanning. *Professional Surveyor*.

Zhang, Z. (1998). A flexible new technique for camera calibration. *IEEE Transactions on Pattern Analysis and Machine Intelligence, 22*, 1330-1334.

Appendix A – Original Method DES Model Variable Definitions

Variable Name	Definition
SubStrcPiersRdy	Substructure piers are built
ConcrtStlRdy	Materials (steel or concrete) for piers are ready for install
GirdersBuilt	Steel or concrete girders are constructed
DeckPansRdy	Deck pans are ready for install
GrdrsRdyForPans	Girders are built and ready for pans to be installed
DeckPansAdded	Installation of deck pans on girders
DeckDrainsRdy	Deck drains are prepared and ready for installation
GrdrsRdyForDrns	Girders have pans and are ready for drains to be installed
DrnsMatsAdded	Drains and rebar mats are placed in the girders
MatsRdyForPlks	Rebar mats are ready for planks to be placed
PlanksCutAndRdy	2"x4" and 2"x8" wood planks cut, prepared, and beveled
WdPlankOnMatsA	Wood planks placed at location on rebar mats in Zone A
ZoneAComplete	All planks placed in Zone A
WdPlankOnMatsB	Wood planks placed at location on rebar mats in Zone B
ZoneBComplete	All planks placed in Zone B
WdPlankOnMatsC	Wood planks placed at location on rebar mats in Zone C
BlcktsPreparedA	Blockouts cut to length, wrapped in plastic, and coated with silicone for Zone A
BlcktsPreparedB	Blockouts cut to length, wrapped in plastic, and coated with silicone for Zone B
BlcktsPreparedC	Blockouts cut to length, wrapped in plastic, and coated with silicone for Zone C
BlcktLocRdyA	Locations marked and ready for blockout installation in Zone A
BlcktLocRdyB	Locations marked and ready for blockout installation in Zone B
BlcktLocRdyC	Locations marked and ready for blockout installation in Zone C
BlcktsPlcdSetA	Blockouts placed and set at proper location in Zone A
BlcktsPlcdSetB	Blockouts placed and set at proper location in Zone B
BlcktsPlcdSetC	Blockouts placed and set at proper location in Zone C
PlnkIdleForConc	Planks prepared and waiting for concrete placement
ConcRdy	Concrete batched and ready for placement
ConcPlaced	Placement of concrete
ConcRdyForFnsh	Concrete placed and ready for finish
ConcFnshrsIdle	Concrete finishers prepared for finishing
ConcFnshByHand	Concrete finished by hand
PlksRdyForRmvl	Planks ready for removal from the concrete
CureTime	Time required to cure the concrete
ConcreteCured	Concrete cured to specified level
PlksLftdFrmCncA	Wood planks lifted from concrete in Zone A
PlksLftdFrmCncB	Wood planks lifted from concrete in Zone B
PlksLftdFrmCncC	Wood planks lifted from concrete in Zone C
AComplete	All wood planks lifted from concrete in Zone A
BComplete	All wood planks lifted from concrete in Zone B
BlcktRdy4MsnBtA	Blockouts ready for drill out with masonry bit in Zone A
BlcktRdy4MsnBtB	Blockouts ready for drill out with masonry bit in Zone B
BlcktRdy4MsnBtC	Blockouts ready for drill out with masonry bit in Zone C
CrpCrewIdle	Carpenter crew ready to drill out blockouts
BlcktsDrlMsnBtA	Blockouts drilled out with a 1.5" masonry bit in Zone A
BlcktsDrlMsnBtB	Blockouts drilled out with a 1.5" masonry bit in Zone B

Variable Name	Definition
BlcktsDrlMsnBtC	Blockouts drilled out with a 1.5" masonry bit in Zone C
InstllPlntStrps	Plinth stirrups to be installed by others

Appendix B – Original Method DES Model Detailed Results

Stroboscope Model Wood Blockout062711.vsd (379086016)
** Calculated results after simulation **

Run	Time	Run	Time	Run	Time
1	112.739336	36	112.657281	71	115.795159
2	114.732284	37	119.210154	72	113.904728
3	115.210341	38	118.306613	73	115.386474
4	115.399775	39	114.928826	74	120.463669
5	118.064766	40	116.084642	75	117.772528
6	119.655011	41	116.413821	76	115.586426
7	114.891779	42	116.964446	77	117.32054
8	109.149721	43	117.283682	78	117.314057
9	112.928105	44	113.659723	80	114.140928
10	123.338288	45	113.952838	81	117.865258
11	112.859955	46	118.472509	82	115.184913
12	113.636987	47	115.813757	83	116.618758
13	115.884091	48	117.086845	84	114.3891
14	116.637424	49	111.521948	85	116.41791
15	116.817183	50	116.026288	86	113.597128
16	115.568706	51	113.908466	87	114.027644
17	111.188738	52	117.655227	88	118.440576
18	111.234569	53	116.317761	89	114.186929
19	117.641332	54	114.828977	90	117.56379
20	110.280375	55	118.396645	91	117.065162
21	116.476981	56	116.055645	92	114.652431
22	114.166278	57	115.329849	93	113.693564
23	115.943048	58	114.357038	94	110.604022
24	116.455815	59	109.664097	95	118.34676
25	111.528682	60	110.525568	96	110.038316
26	121.495206	61	113.005376	97	116.100455
27	115.154709	62	111.300145	98	117.47368
28	113.338811	63	119.88507	99	121.628479
29	113.486762	64	117.639524	100	121.757858
30	111.398723	65	115.474629		
31	115.330127	66	109.992529		
32	114.485419	67	113.466482		
33	112.91469	68	118.759435		
34	113.258141	69	118.070969		
35	112.240729	70	119.840905		
Average	115.45529				
Std Dev	2.89711442				
Minimum	109.149721				
Maximum	123.338288				
Execution Time = 2.636 s					

Appendix C – 3D Imaging Method DES Model Variable Definitions

Variable Name	Definition
SubStrcPiersRdy	Substructure piers are built
ConcrtStlRdy	Materials (steel or concrete) for piers are ready for install
GirdersBuilt	Steel or concrete girders are constructed
DeckPansRdy	Deck pans are ready for install
GrdrsRdyForPans	Girders are built and ready for pans to be installed
DeckPansAdded	Installation of deck pans on girders
DeckDrainsRdy	Deck drains are prepared and ready for installation
GrdrsRdyForDrns	Girders have pans and are ready for drains to be installed
DrnsMatsAdded	Drains and rebar mats are placed in the girders
TrgtsAvailable	Target available for measurement
CoordSysDefined	Define the site coordinate system for layout
ImgngEqpSetupA	Setup imaging equipment in Zone A
ImgngEqpSetupB	Setup imaging equipment in Zone B
ImgngEqpSetupC	Setup imaging equipment in Zone C
MsrRgstrtnTrgtA	Measure registration targets in Zone A
MsrRgstrtnTrgtB	Measure registration targets in Zone B
MsrRgstrtnTrgtC	Measure registration targets in Zone C
AcquireDataA	Imaging equipment acquires data in Zone A
AcquireDataB	Imaging equipment acquires data in Zone B
AcquireDataC	Imaging equipment acquires data in Zone C
DataCollected	Data collected from equipment and ready for processing
ZoneAImageCompl	Imaging from Zone A complete to release work downstream
ZoneBImageCompl	Imaging from Zone B complete to release work downstream
ZoneCImageCompl	Imaging from Zone C complete to release work downstream
WdPlankOnMatsA	Wood planks placed at location on rebar mats in Zone A
WdPlankOnMatsB	Wood planks placed at location on rebar mats in Zone B
WdPlankOnMatsC	Wood planks placed at location on rebar mats in Zone C
PlankAComplete	All wood planks placed at location in Zone A
PlankBComplete	All wood planks placed at location in Zone B
ProcessData	Post process data for analysis
DataProcessRdy	Equipment ready to process data
FndLctnsFrmData	Analyze data and extract rebar locations
DtrmnDrllLctns	Determine drill locations from data
PlnkIdleForConc	Planks and forms waiting for concrete placement
ConcRdy	Ready-mix concrete ready for placement
ConcPlaced	Placement of fresh concrete
ConcRdyForFnsh	Concrete placed in form and ready for finishing
ConcFnshrsIdle	Concrete finishers ready to finish the concrete
ConcFnshByHand	Concrete finished by hand
ConcCured	Concrete cure time
ConcreteCured	Concrete cured to specified level
CrpCrewIdle	Carpenter crew ready to lift planks
PlksLftdFrmCncA	Wood planks lifted from cured concrete in Zone A
PlksLftdFrmCncB	Wood planks lifted from cured concrete in Zone B

Variable Name	Definition
PlksLftdFrmCncC	Wood planks lifted from cured concrete in Zone C
LiftAComplete	All wood planks lifted from concrete in Zone A
LiftBComplete	All wood planks lifted from concrete in Zone B
LctnMrkdOnDeckA	Mark locations for drilling on concrete deck in Zone A
LctnMrkdOnDeckB	Mark locations for drilling on concrete deck in Zone B
LctnMrkdOnDeckC	Mark locations for drilling on concrete deck in Zone C
LctnsDrlldOutA	Locations for dowel bars drilled out of concrete deck in Zone A
LctnsDrlldOutB	Locations for dowel bars drilled out of concrete deck in Zone B
LctnsDrlldOutC	Locations for dowel bars drilled out of concrete deck in Zone C
InstllPlntStrps	Plinth stirrups installed

Appendix D – 3D Imaging Method DES Model Detailed Results

Stroboscope Model 3D Imaging062711.vsd (2128867904)
** Calculated results after simulation **

Run	Time	Run	Time	Run	Time
1	111.531308	36	115.38654	71	111.576123
2	113.538321	37	115.244002	72	109.78071
3	108.381203	38	109.257344	73	106.921773
4	107.563992	39	108.924679	74	113.020954
5	113.317412	40	107.751987	75	110.824677
6	107.904106	41	109.219645	76	105.739245
7	111.065636	42	111.586718	77	105.840171
8	107.820183	43	104.704886	78	111.01354
9	108.575478	44	115.593818	79	107.637362
10	110.52329	45	113.511223	80	109.932878
11	109.92525	46	111.124898	81	109.456153
12	112.272405	47	109.057267	82	113.000171
13	107.815178	48	115.107664	83	110.190399
14	108.513354	49	111.347674	84	106.032673
15	111.576008	50	110.281069	85	111.59375
16	110.488901	51	107.123951	86	111.54743
17	107.531666	52	111.922755	87	113.187886
18	111.843827	53	111.21423	88	106.911036
19	110.715614	54	109.773642	89	111.212273
20	109.122995	55	113.19829	90	111.072454
21	111.880755	56	114.438051	91	108.508338
22	106.105421	57	104.563378	92	110.934125
23	106.565769	58	108.892706	93	107.609447
24	112.882559	59	109.06692	94	107.094522
25	108.50519	60	110.489274	95	114.609995
26	112.23953	61	110.583211	96	110.651569
27	110.074044	62	106.837402	97	113.913195
28	109.804371	63	105.852495	98	106.868951
29	115.999775	64	112.557361	99	105.066579
30	110.639707	65	107.951833	100	108.995658
31	109.746336	66	110.04081		
32	108.630525	67	109.866629		
33	113.627551	68	107.636611		
34	110.713597	69	113.157541		
35	108.055448	70	109.830062		
Average	110.093373				
Std Dev	2.6144283				
Minimum	104.563378				
Maximum	115.999775				
Execution Time = 2.823 s					

130

www.ingramcontent.com/pod-product-compliance
Lightning Source LLC
Chambersburg PA
CBHW080253180526
45167CB00006B/2516